Never Lost Again

NEVER LOST AGAIN

The Google Mapping Revolution That Sparked New Industries and Augmented Our Reality

BILL KILDAY

HARPER
BUSINESS

An Imprint of HarperCollins*Publishers*

HarperCollins books may be purchased for educational, business, or sales promotional use. For information, please email the Special Markets Department at SPsales@harpercollins.com.

FIRST EDITION

Designed by Bonni Leon-Berman

Art courtesy of Shutterstock

Library of Congress Cataloging-in-Publication Data has been applied for.

ISBN 978-0-06-267304-6

18 19 20 21 22 LSC 10 9 8 7 6 5 4 3 2 1

For Shelley.

The S in GPS.

Once a photograph of the Earth, taken from the outside, is available . . . a new idea as powerful as any in history will be let loose.

—SIR FRED HOYLE, BRITISH ASTRONOMER (1948)

CONTENTS

INTRODUCTION

Starting Point

Do you remember the last time you were lost? I mean, truly and totally lost?

Personally, I have not been *that* lost in some time. I'd have to go all the way back to the early 2000s—to when I was living in Boston, Massachusetts. One cold winter evening, I was driving home from a Tuesday-night pickup basketball game in Brookline. I was exhausted from three hours of rough-and-tumble play with some South Boston Irish guys. Somehow I got turned around on my way home, a route that I had already driven more than a dozen times. I could see where I needed to be—just across the Charles River—but I couldn't figure out how to get there.

My wife, Shelley, at home with a screaming baby, had already called twice. "Where are you?" In frustration, I pounded my fist against my car's dashboard, yelling at nobody but myself, while driving five miles in the wrong direction down Route 2, looking for the next roundabout. Or maybe it was 3A?

From 2000 to 2003, I lived in Boston—and I was frequently turned around. The city was merciless to a transplanted Texan: It was like a foreign language. The locals seemed to take pride in the missing signage, serpentine streets, and roundabouts. You needed to solve a math equation to navigate some intersections. Throw in the Big Dig—the $15 billion public works project that aimed to reroute four miles of Interstate 93 directly *under* the city of Boston—and yeah, I was lost. A lot.

I'm not sure I ever did figure it out. "Boston," you see, is actually a twisted collection of cities. If you place them on a clock, you would start at twelve o'clock with Chelsea, then go clockwise through Revere, Boston (proper), Brookline, Brighton, Newton, Belmont, Cambridge, Somerville, and Charlestown, with a dozen or so smaller cities filling in the gaps. These cities were all at one point independent municipalities, each established in the early to mid-1600s, and connected by a complex system of horse trails.

The way that each of the different cities named trails was quite simple: They named it according to the trail's destination. For example, if you lived in Brighton in the 1600s and wanted to ride to Cambridge, you took a horse trail that was marked *Cambridge*. Similarly, if you lived in Boston and wanted to ride to Cambridge, you also followed a horse trail marked *Cambridge*.

Fast-forward to modern-day Boston, and you have—I'm not making this up—at least seven different Cambridge Streets: The names of the now-paved trails in the Boston metro area were often inherited from the names of the original horse trails. I can remember driving along Cambridge Street one day, stopping at a red light, and glancing up at the cross street I was intersecting: It was Cambridge Street!

It took me two and a half years to figure out why, in the Boston metro area, there appeared to be duplicates, quadruplets, septuplets of every street name (assuming you could find a sign). One day I asked my landlord why the roads were still set up in this way, and he answered, "So the Yankee fans can't find their way to Fenway when they drive their cahhs up from New York."

Well, that made me feel better.

Ten years later, in 2010, I was on a vacation in Portland, Oregon. I walked through Pioneer Courthouse Square with my family; the landmark public space was alive with people and activity on a crisp summer evening. Following Google Maps on my iPhone, I navigated the half mile down Yamhill Street to a restaurant called Luc Lac Vietnamese Kitchen. My eight-year-old daughter Isabel asked me, "Daddy Bill, what did people *do* before Google Maps?"

I thought back to those Boston days. To the nights left to chance and happenstance, like a Yankee fan in Boston's North End before Yelp and OpenTable, before the iPhone and Google Maps. To a world in which I would have never known about a restaurant like Luc Lac (4.5 stars!), and if I had, would have had to stop strangers for directions three different times. I thought back to all those miscalculated turns and premature roundabout exits. And all those restaurants I walked into without (shudder) reading reviews and those hotels booked without (horror) looking at the street-view imagery of the block where they resided.

I looked up from my Google Maps app, slid my iPhone into my vest pocket, smiled at Shelley, and tried to answer.

"We got lost a lot, sweetie."

Before 2004, we printed out maps from MapQuest and crammed poorly folded AAA maps in our glove compartments or under our seats. We stopped at gas stations and asked strangers behind bulletproof plexiglass for directions. We asked hotel concierges for dubious restaurant recommendations. And when on vacation, we rented condos that stretched the definition of *beachfront*. We studied confusing subway maps (Green

Line not E). We got lost, we settled for mediocre meals, and we pounded on our dashboards.

But by 2010, a navigation and mapping technology had changed everything forever. And I was there to witness its creation—and played my small part.

A friend of mine from the team, Michael Jones, put it to me this way. "Think about it: Two hundred thousand years of human beings on Earth—and we are the very last generation to ever know what it was like to be lost. And no one after us, no generation to come will ever be lost in the way that every generation before was. All over the world today, people are landing in New York or Tokyo or Cairo or even the Democratic Republic of Congo. Anywhere, they are stepping off of planes in a foreign land, in a place that they've never been, and guess what, they know where they are. They can find their hotel. Or a restaurant. Or a friend's house. Or an office for a business meeting. Or whatever."

He paused, looked at me, and added, "And we did that. You and me and everyone that worked at Keyhole and Where2Tech and the Google Maps team. We fixed that!"

Back in Portland, I opened the door to Luc Lac on that summer evening, knowing already it was the perfect restaurant. I knew it was open, I had prescreened the menu, and I had navigated there with ease, thanks to the Google Maps app on my iPhone. I knew it wasn't going to be too fancy or too expensive.

Shelley passed me with a knowing grin on her face. She remembered our Boston days. She remembered how it all came to be—the technology that changed our lives forever. And now she was also one of its billion monthly users.

Keyhole started, barely, in 1999. It ran out of cash, essentially, in 2002. It was resuscitated by CNN and In-Q-Tel (the venture capital arm of the CIA) in 2003. And by 2004, it was bought by another five-year-old company.

That company was Google.

Google bought two mapping companies that year: a four-person team working out of an apartment in Sydney that had not yet even incorporated as a business; and a twenty-nine-person company in California, called Keyhole, for which I served as marketing director.

In the fall of 2004, Google combined those two teams with a small group of existing Google employees in Building 41 in the Googleplex in Mountain View, California, gave them zero direction, unlimited resources, and presented the teams with a secret problem: *Twenty-five percent* of all queries being typed into the simple white Google search box were looking for a map.

And guess what? Google had no map.

Searches like "best margarita, Austin," "hotel in New Orleans," "heart attack near zip code 78636" were representing a significant share of all Google traffic. As late as January of 2005, if you entered those searches looking for a location or directions on the Google home page, Google returned a list of ten website links, with a *Gooooogle next* button on the bottom, leaving its users still in need of a map and directions. Good luck with that heart attack.

Six years later, Google's mapping products—run by several key members of the former Keyhole team—had one billion monthly active users and became the number one consumer mapping service worldwide. From zero users to a billion users monthly. In six years.

And our work ended up sparking entire industries: Yelp, OpenTable, Zillow, Priceline, Uber. And hundreds of other services whose economic prospects were viable only because someone else had done much of the heavy lifting. Someone else created the ultimate base map, the blank sheet of paper on top of which whole new businesses could be drawn. And then gave that base map away via the freely available Google Maps API (or application programming interface).

In 2007, Google squeezed all of those maps and services into your pocket, and Google Maps became the killer app of Apple's killer new device: the iPhone, with Steve Jobs personally demanding the inclusion and implementation of Google Maps. Android phones soon followed.

And finally in 2008, Google redoubled its outrageous investments in mapping with two even more audacious moonshot projects: the Street View project and Project Ground Truth, ultimately setting into motion the future of self-driving cars.

How did we get there? Honestly, when I think back on the Keyhole journey, where it all started, I am still in awe of how it happened. I mean, I was there. I was more than along for the ride. I had my small role in its success.

More than once I have thought, *There's no way Keyhole should have survived, the company could have failed so many times, we were so lucky, and so many things fell into place for us, and it could never happen again.*

But I know my Keyhole colleagues well. Looking back now, I know there was no way that we would have failed: Whatever the obstacle, whatever the missed turns, we would have found our path. After all, we had an Ace up our sleeve. He was going to figure out a way, no matter what.

WHERE ARE WE GOING?

The Start-up Years

Chapter 1

THE SUPERMAN THING

On an unseasonably warm day in the spring of 1999, I got a call at work from an old college friend, John Hanke. I was working as the marketing director for the *Austin American-Statesman* website. "Hey, dude, I'm in Austin," he said. "I've got something to show you. Can I come by your house?" At the time, John was working for a new start-up in Silicon Valley, but didn't want to tell me about the project on the phone. I pressed him for more details, but he insisted on coming by that night. "It's really something you need to see in person."

John and I had been friends for fifteen years, having met the Sunday before the first day of classes our freshman year at the University of Texas in 1985. I was assigned a room in Jester Residence Hall. The residential hall occupied a full city block; at the time, it was the largest dormitory in North America, with 3,200 students, and had its own zip code. Its generic rooms and endless fluorescent-lit hallways were almost prisonlike and not exactly hospitable for freshmen away from home for the first time. That evening I signed up on the sheet outside my RA's door to go to dinner with a group of girls from Kinsolving, the all-female dorm on the other side of campus.

It is worth noting that only undergrads with nowhere else

to eat signed up for such an outing. The dorm's cafeterias were closed on Sunday evenings, and we were left to our own devices to find sustenance. For many freshmen, this meant dinner at their fraternity or sorority house, assuming of course that you had the means to join a fraternity or a sorority. And even if you weren't in the Greek system, you should at least be able to find a friend and order pizza together.

Signing up on that RA's list was taking a risk in a way, socially speaking. It was essentially adding your name to a list that could have been entitled "List of Students with No Money or Friends." My roommate, Kevin Brown from San Marcos, Texas, was an accomplished trumpet player, had joined the Longhorn Band, and was already busy with his new friends from the band. After seeing five other students on the list, I added my name.

I was concerned as my floormates gathered outside my RA's door at the appointed time: a foreign exchange electrical engineering major from Korea; a burly kid from Harlingen; me, a pimply-faced, six-foot-three-inch string bean. And this quiet, serious kid who lived eight doors down from my room and whom I hadn't quite figured out. He was a Texas-tan, good-looking guy with a medium build, and appeared to be working on his non-ironic mustache. Maybe a West Texas version of Charlie Sheen. Through the open door of his room, I had noticed a funny-looking personal computer. He was the only student on our hall with his own computer.

I seriously considered ditching dinner, but I had been the one to recruit the kid from down the hall, so I was stuck. Surveying our crew that night, I was worried about our floor's ability to represent. You see, I knew what awaited us on the other side of campus: Kinsolving was home to six hundred freshman coeds, and I had gotten a job as the cafeteria's salad bar guy there.

Somewhere along that hot summer night's long trek across the forty-acre campus from Jester to Kinsolving, I found myself walking with the reserved young man with the mustache.

"So, what's your major?" I asked.

"I'm Plan 2."

"You're Plan 2!?"

"Why is that so surprising to you?"

"Oh no, it's just that I saw one of those T-shirts on campus earlier this week. You know, the one that reads, *I haven't declared a major, but I Plan 2.* I thought it was genius."

"Yeah, they gave us those at our orientation, but I haven't worn mine," he said with a laugh.

"Why not?" I asked, half expecting him to say it was the wrong size or color.

"It's a little showy, don't you think?"

For most in the program, this showy shirt was worn proudly on campus because it meant that you were a cut above, part of a select academic pedigree worthy of a more rigorous and independent curriculum that the university crafted for the valedictorians and National Merit Scholarship finalists in the crowd. It was common knowledge that many of these students could have gone to Princeton, Harvard, or Stanford, but had decided on the unique interdisciplinary program of UT's Plan 2.

I had obviously underestimated this guy.

"Where are you from?" I asked.

"A little town out in West Texas. How about you?"

"Houston," I replied. "Austin must be a change?"

"Well, Jester *is* about three times as big as my hometown," he shared.

"Ha!" I guffawed in amazement and then said to the students walking a few paces in front of us, "Hey, guys, our *dorm* is three times as big as this guy's hometown!" John didn't seem to find the humor in this.

I remember almost nothing of the girls we ended up having dinner with at Conans Pizza that night; the boys stuck to one end of the table and the girls to the other. I do remember talking with John more, making plans to attend the mass at the University Catholic Center later that night after discovering we were both guilty Catholics, and even ribbing him about the mustache.

"How long have you been working on that thing, John?"

"I've had it about a year," he admitted. "It helps when I go try to buy beer."

"I would have guessed about two weeks!" I said with a laugh.

He had taken a bite of pizza, so he smiled and, as a college male signal of friendship, shot me the finger.

Beer or no beer, when I saw John on the way to class the next day, the mustache was gone—but not forgotten, because it was forever immortalized on his student ID, which you basically needed to pull out four times a day for four years. It was an endless source of entertainment for me and embarrassment for him (though I'm pretty sure he still has it).

John was from Cross Plains, Texas (population 893), and it seemed to me that his hometown was something he did not want to dwell on. It's not that he was embarrassed by his rural roots—quite the contrary. He didn't hide his background: His father, Joe, was a small-scale cattle rancher and town postmaster; his mother, Era Lee, was active in the Catholic church and the local Chamber of Commerce. Cross Plains represented a point of pride for John—and it was something he protected. I learned quickly that it was okay for John to tell someone about the one-stoplight town, the Friday-night social scene centered around the Dairy Queen, 4-H livestock shows, and the 2A football team, but it was *not okay* for anyone else to talk or joke about it. Curiously, the town's most notable inhabitant had been Robert E. Howard, an author who out of the desolation of West Texas somehow created the fantastic new worlds of the Conan the Barbarian book series in the 1920s and 1930s.

By contrast, I had grown up in Houston in what, by almost all accounts, would be considered a decidedly middle-class and unremarkable upbringing, save for a few details: I was the youngest of eight kids and have six older sisters. I was a surprise kid, with seven years between me and the sibling above me. My father, an affable Bostonian ad man for the oil business, passed away in 1983 when I was a junior in high school. All of this made me a hardworking kid with multiple odd jobs in order to make my way through UT. But to John, I was from the big city; Houston was downright cosmopolitan by comparison to Cross Plains. We quickly discovered that we shared a variety of interests—from politics (both progressive) to sports (we attended UT football games together) to live music to our Catholic upbringings.

By the end of our first semester, John and I were close friends. So much so that over winter break, John, my roommate Kevin Brown, and I took a road trip to go skiing in Winter Park, Colorado. It was a first for John and me.

On the way to Colorado, we spent the night in Cross Plains,

about a half hour north of Abilene, playing basketball with his friends at the high school, hitting the local Dairy Queen, and meeting his parents and his older sister, Paula. While it is common for parents to be proud of their overachieving children, in the case of Joe and Era Lee Hanke, their pride in John, who was adopted, was palpable. They were so pleased that he had brought his college friends home for a visit. His father measured Kevin and me up, speaking with a thick drawl. "You boys ever been out this way before?" His neck was weathered from years working Red Angus cattle and other livestock under the West Texas sun.

I got the sense that John's parents—and many others in the town, for that matter—were perplexed by John's ambition and drive: the high school valedictorian of his class of twenty-two teenagers; the student body president; the National Merit finalist. He began coding his own shareware games and selling them through personal computer magazines. Under the mentorship of his math teacher, John participated in a computer programming competition at Baylor University; his team placed third in the state. To his family and friends, he was almost otherworldly.

His high school English teacher, a salt-of-the-earth ranch woman named Clara Nell Spencer, first noted the spark in John. Mrs. Spencer reached out to her sister, who was a guidance counselor at Westlake High School in Austin. Together, the sisters conspired and helped John apply to UT.

John seemed embarrassed by it all. The next morning we got up early. Joe checked the car's oil and tire pressure. He unfolded the gas-station map on the hood of the car and pointed out the best route. As we said goodbye to his parents, I had a hunch that they weren't expecting him to marry his high school sweetheart and to take over the family farm. Joe and Era Lee were more than okay with this. They—and many others in the town—looked forward to seeing where their favorite son might end up.

Two days later we were on the slopes. It was brutally cold—with a windchill of seventeen below zero. Since neither of us had the money for ski lessons, we relied on Kevin, who was a veteran skier, for instructions. On our first chairlift ride, I stumbled through the line, planting my ski pole as the chair swung through, snapping the pole in two. I cautiously skied off at the top of the mountain, with my broken pole, and waited for my next set of instructions.

John, on the other hand, skied right past Kevin and me, already a bit out of control. Since this was our first trip up the mountain, Kevin had not given us any important ski tips, including how to stop. Kevin and I stood there. First with curiosity, then amazement, then in horror as John pointed his skis straight down the mountain.

"Turn, turn, *turn*!" Kevin shouted as John picked up speed. It looked like he was at least "trying" to turn. But instead of turning, he lifted one ski and then the other, all the while pointing straight down the mountain. His spectacular run concluded not merely with a fall, but with an explosion of snow, skis, poles, and limbs. It took me about ten minutes of careful maneuvering to cover what John had "skied" in the space of seventeen seconds.

"Dude, what took you so long?" he asked.

This was John. He was willing to take risks—and it would serve him well. I came to know him as an intense, hardworking, and ambitious guy. Somehow out of Cross Plains, he adapted this tenacious approach to life and work. Maybe it was the hardscrabble ranching way of life? Or maybe it was some of the feelings that came with being adopted? I like to think that I have been a good friend to him over the years because I wasn't quite so intense and ambitious. To this day, I still try to make him laugh every once in a while, and slow down and take it easy.

During our college years, John pulled me toward student union leadership committees, better grades, and Sunday mass. I drew John to spring breaks at South Padre Island, intramural sports, and live music at Liberty Lunch. There was one harrowing episode in which John got buried in a mosh pit of a Replacements concert, but I managed to pull him up from the floor by his collar.

After graduation, John took a job with the US Foreign Service in Burma (now Myanmar). Now if you drilled a hole from Cross Plains, Texas, straight through the Earth, you'd come out very close to Burma. I'm not sure that was a coincidence. You couldn't have gone any farther away from Texas. He was drawn to the allure of world travel and experiences. At the same time, it seemed to me that John became more private and spoke very little about his work, as if there was another protective shell surrounding him. During these years, I stayed in Austin because I had been accepted into the dual graduate program of the Lyndon B. Johnson School of Public Affairs and the McCombs School of Business.

In early 1991, John called me late one night on a static-filled phone line to tell me that he had proposed (over the same static-filled line) to his bride-to-be, Holly Hayes, and he asked me to be his best man. Holly and John had met in Washington, D.C., where they were both working for the State Department. The connection was so bad I had to ask him, "Are you *sure* she said yes?" The black-tie wedding in McLean, Virginia, was an interesting mix of international travelers, foreign service officers, and Cross Plains family and friends.

In the summer of 1993, our relationship began to evolve from friendship to doing business together. That July, I invited John and a third friend, Carl Townsend, for an epic nine-day California road trip: from Tijuana to Tahoe up Highway 1. I had just graduated from McCombs and John was starting at Cal's Haas School of Business. I remember throwing my bag into the back seat of John's rented convertible Mustang and spotting the neon-hatch-marked spine of a *Wired* magazine. "Hey, you read *Wired* magazine, too?" I asked.

"You read *Wired*?" John was equally surprised as I pulled my copy out of my backpack.

"Oh, yeah, I'm working on interactive marketing for this ad agency. We're building CD-ROMs and websites for all of our clients now, including Dell," I said. Back in Austin, I was helping build the very first Dell website.

"I didn't know you were working on websites now. Dell? That is huge," John said. He knew the Michael Dell story well. Dell was an infamous Houston kid who had been kicked out of our UT dorm in 1984 for building, selling, and servicing computers out of his room.

Now, both of us owning a copy of *Wired* might not sound like that much of a coincidence, but in the summer of 1993, *Wired* magazine was a relatively obscure publication, covering the initial wave of tech entrepreneurs who were rushing into the golden hills of Silicon Valley to claim their Internet riches. Keep in mind, the first dial-up access to the Internet had emerged only in 1992, and the Mosaic graphical web browser, created at the University of Illinois at Urbana–Champaign, had just been released in 1993. Its inventor, Marc Andreessen, graduated from Illinois and formed Netscape with Jim Clark in 1994. All of this set the stage for the swift commercialization of the Internet. There couldn't been a better time to start a career in technology.

For all of our mutual interests, John and I had not talked much about technology or business. But during the trip we conspired endlessly about what it might all mean—this whole commercialized Internet thing. Within the year, while still in school he started an Internet gaming company with two other Berkeley Haas students, and I traveled to California to help him on various marketing projects, such as websites, banner ads, and marketing literature. This continued throughout 2000 as I managed to squeeze in projects for John while I worked full-time at the ad agency and later at the *Austin American-Statesman*.

So when John called that warm spring day in 1999 and told me about the demo he wanted to show me, I fully expected to be helping him out with another marketing-related project. Around nine o'clock that evening, John rolled up with a software engineer named Brian McClendon in front of my house. John now sported a buzz cut rather than his former swoop of brown hair. And he wore the standard uniform of the young Silicon Valley start-up CEO: a blue blazer, T-shirt, jeans, and a messenger bag slung over his shoulder. Brian was barefoot and wore bright yellow cargo shorts. Together, they carefully lugged in a huge Dell PowerEdge server packaged in a Pelican-brand, hard protective case normally reserved for expensive camera equipment. It was clear that the prospects of their start-up company resided within its confines. My dog, Penny, trailed the pair, violently wagging her tail.

Brian immediately sized me up. "Six four, two hundred and fifteen?" he asked.

"Uh, yeah."

"How tall are your ceilings?" Brian asked. "Ten feet," I responded, and he leapt up and touched the living room ceiling (barely). "Can you dunk?" he asked, collecting the loose change that had spilled from his pockets. "For about six months I could," I replied. "Sounds about right," Brian said with a laugh. As it turned out, we were the exact same height and weight.

I introduced Brian to Shelley, who was at that point my fiancée. A native of Los Angeles, Shelley was working in urban planning for the city after having graduated from the LBJ School. We had been introduced by friends at a bar in downtown Austin.

Soon John connected the server to a monitor and it was heating up in my spare bedroom. Penny sniffed at the machine. "Okay, come on in," John said to Shelley and me. "There's a good chance it will crash, but I want to try to show you something."

John was ready. His face filled with excitement. The keyboard rested on his lap, with the giant server humming loudly next to him. *This has to be a heck of a demo to carry this enormous server around Texas,* I thought, somewhat skeptical.

On the screen floated a perfectly detailed rendition of Planet Earth. *Oh,* I thought, *there's that photo.* You know the one. It was on the cover of every junior high science book. Known simply as the Blue Marble, this classic image was captured on December 7, 1972, by the crew of the *Apollo 17* spacecraft five hours into a lunar mission. This one spectacular photo is widely credited with jump-starting the modern-day environmental movement.

The Earth seemed to be rotating. Moving somehow. "What's your address again?" John asked. "It's 465 Joe Sayers, Austin, Texas," I said as he typed.

That's curious, I thought. *The photo on the screen. The Earth. It is moving! Maybe it's a QuickTime movie animation of some sort? But QuickTime movies aren't interactive in any way, they are linear. Why would John need my address if this were a movie and . . . Wait, how is this thing zooming in? What the hell is happening here!?*

These thoughts raced through my head during the fifteen seconds that we zoomed from outer space down to what was clearly my house. I recognized my standing seam metal roof, the neighbor's trampoline in their backyard, my red Ford Explorer parked out front, and the Arroyo Seco running behind the house.

"Holy shit!" I exclaimed.

"Pretty amazing, isn't it?" John said.

I was awestruck. I high-fived John and Brian, yelling for John to type in the addresses of my childhood haunts. "Try my mom's house, 708 Atwell, Bellaire, Texas," I said. "Holy Ghost Catholic School. How about Bellaire Little League field?" As I looked at the screen, I could see where I hit a ball from home plate to the dead center-field wall, mere inches from a home run. I could almost hear the ball slap against the wall. Nothing could have transported me back to that moment when I was eleven years old more than seeing it there on the computer monitor, somehow photographed from space.

John then flew over and into the Grand Canyon, and tilted the viewing angle to show the 3-D terrain, rotating around and piloting over the South Rim down into the canyon like a bird in flight. There were the dramatic layers of pinks, oranges, and browns. I put my hand on John's shoulder. My knees almost buckled. I couldn't believe what I was seeing.

Shelley said that it reminded her of something out of the movie *Enemy of the State*. While she was equally impressed, Shelley also stayed grounded. She is honest. And Shelley had a question. And it wasn't an easy one. Answering that question would in fact take the next several years of many lives and cost many millions of dollars. "But what do you *do* with it?"

John and Brian may have had a plan as to what to do with this technology, but it wasn't clear to me. It was truly an incredible demo, but an incredible demo does not a company make.

"We're going to close a seed round of $10 million in the next few weeks," John said with the swagger of presumption and confidence that comes with the territory of being a Silicon Valley start-up CEO. Apparently at least one venture capitalist could see the potential of the demo.

"What's the name of the company?" I asked.

"Keyhole," John said, adding, "for now at least. That's actually why we are in Austin. Ultimately we want to launch as earth.com."

John and Brian had come to town to meet with the owner of the earth.com domain. He was an IBM employee based in Austin who had been smart enough to grab the earth.com domain in 1992. His price for the domain: a million dollars. While Brian gave me the rundown of the meeting, John didn't want to talk about it. They both hoped that a demo might persuade him to ask less, or at least consider an exchange of some equity instead. Despite the demo, he was not budging off his price. We spent the hour in my living room relaxing over a beer, debating the value of earth.com, and carefully packing up the EarthServer, as they called it, for the return flight to California.

As they headed out the door, John took another run at trying to convince Shelley of the economic merits of what she had described as "the superman thing." Unconvinced, Shelley had let John down gently, telling him with a motherly pat on the shoulder that she didn't really understand technology, and that she was certain there were innumerable industries that would line up to pay for such a service.

What John didn't share, however, was that in the spring of 2000, the appetite of venture capitalists for consumer start-ups without a clear path to revenue was beginning to cool. The runaway train that began in 1993 with the introduction of the first web browser had fueled unprecedented speculation, valuations, and spending. But now, in early 2000, up and down Highway 101 in Silicon Valley, the formerly high-flying dot-coms of the late 1990s were facing a harsh new reality: a reality not based on monthly active user vanity metrics, but instead based on profits and losses. Investors wanted to see real revenue before investing.

If it hadn't closed already, John's window for raising investment capital was closing quickly. The demo may have been extraordinary, but the timing could not have been worse: There would be no $10 million.

Chapter 2

POWERS OF TEN

"We're going to close on $5 million in the next couple of weeks. You should plan a trip out to Mountain View," John had said to me over the phone almost every month since the demo at my house in Austin.

Throughout the spring of 2000, I helped John with a few small marketing tasks for his company even though he had not closed on any investment capital. At the time, I was still working on the website for Austin's newspaper. My inaugural visit to the Mountain View office on California Street in March of 2000 was also my first experience with the classic Silicon Valley work environment: the gray cubicles, makeshift sleeping quarters under desks, stacks and stacks of old pizza boxes, and empty Mountain Dew cans strewn everywhere. A black Lab roamed the office, and children arrived for late-night pizza with their overworked parents.

I use the word "office" liberally here. The Keyhole team shared a single oversized cube stuffed in the corner of another start-up: Intrinsic Graphics, which was founded by Brian McClendon, Michael Jones, and two other software engineers, Chris Tanner and Rémi Arnaud, in 1998. It was difficult to tell the two companies apart because John's start-up had not yet raised investment capital. As a result, an incubator-like relationship had emerged. In fact, the only people getting paid to work on John's new company were being paid by Intrinsic Graphics; John was essentially running a bar tab for software development.

The Intrinsic founders had been colleagues at the famed Silicon Graphics (SGI), which was created by tech innovator Jim Clark in 1982. Silicon Graphics was a pioneer in the field of 3-D graphics and the epicenter of innovation and talent in creating anything from car parts to buildings to virtual worlds in 3-D on computers. In the 1980s and early 1990s, if you wanted to develop the hardware, software, or content that involved anything 3-D, you wanted to be at Silicon Graphics. But by the mid-1990s, SGI's market came under attack from cheaper, powerful Windows-Intel (aka Wintel) workstations from companies like Dell and Hewlett-Packard.

The Intrinsic founders left SGI and soon recruited a core team of some of the best programmers in the highly specialized 3-D visualization space, and set them to work on creating a suite of software tools that could be used by game developers to build interactive 3-D environments. Most of the team had previously worked on high-end 3-D simulator projects: flight simulators, combat simulators, nuclear plant simulators. These projects often cost multiple millions of dollars to develop and ran on dedicated high-end hardware set up in an artificial stage built to re-create the effect of occupying the actual physical environment being simulated. For example, if you were flying an Airbus A320 simulation, you sat in the cockpit with all of its instruments in front of you, with high-resolution computer monitors in the place of the cockpit windows.

The engineers founded Intrinsic Graphics because they recognized that a wave of affordable, super powerful PCs could bring to the desktop computer what was previously available only on multimillion-dollar specialized simulators. They were betting on Moore's law: Gordon Moore was the cofounder of Intel and famously forecasted that the number of transistors on an integrated circuit would double, and would continue to double every two years.

A side project at Intrinsic revolved around a specialized graphical innovation called clipmapping, a patented technique for loading a multi-resolution stack of images and blending them together to create a seamless mosaic. This is how an engineer explained it to me: Clipmapping is a method of clipping a precalculated, optimized sequence of images—or mipmaps—to the subset of geometry being rendered in a 3-D screen scene. Chris Tanner's and Intrinsic's patented clipmapping work determined how to load the *least* amount of data possible and still quickly render realistic 3-D scenes on a screen.

Did you get that? Imagine for a moment that you're standing on a ten-meter platform above an Olympic-size swimming pool. There's a quarter at the deep end of the pool that you need to retrieve. You see it and dive down. In a scene based on Chris's clipmapping, only the water that you dove through would be loaded, not the entire volume of the pool. To you, it looks like you dove into the pool. What you don't know is that only a fraction of the pool's water—or only the water you could see—was loaded. Clipmapping calculated the minimum amount of water to display during your dive, and then showed you only water you could possibly see, not the entire pool.

All of this translated into a much faster visual experience than if the full data of a scene was loaded over the Internet. In early 1999, this technology was being used for flight simulators and video games, but one weekend Michael Jones gathered with Chris and Rémi around his kitchen table and worked together on applying this technology to a new use: a map. A map with its DNA rooted in video games and simulators—a map that would be faster than any other digital map before it.

As inspiration, Michael showed Chris and Rémi the seminal nine-minute film *Powers of Ten* made by architects Charles and Ray Eames in 1977. The cult classic aimed to explain the relative size of things, zooming from a couple picnicking in Chicago's Grant Park to distant space and back again; each zoom level represented another "power of ten" as the camera moved. The film provides a remarkable visual effect and served as the starting point for the experience that the engineers set out to re-create that weekend around Michael's kitchen table. This concept of a digital model of the Earth was an archetypal idea. In 1998 Vice President Al Gore had discussed the idea of a 3-D digital Earth in a speech at an education conference, describing a future where "all the world's citizens could interact with a computer-generated 3-D spinning virtual globe and access vast amounts of scientific and cultural information to help them understand the Earth and its human activities." And in Neal Stephenson's science fiction novel *Snow Crash* (1992), the protagonist Hiro uses "a piece of CIC software called, simply, Earth. It is the user interface that CIC uses to keep track of every bit of spatial information that it owns—all the maps, weather data, architectural plans, and satellite surveillance stuff." SGI had also created a proof-of-concept demo called "Space to Face" that ran on a $2 million SGI Infinite Reality computer. So the idea was not completely new.

But Michael and the team were the first to create a digital model running on something that might actually be accessible to the masses—on a personal computer. (The team used a Dell computer, priced at about $4,000.) The "CTFLY" demo—as Michael soon called their work—was an extraordinary application. In their model, the user zoomed from outer space down to a single high-resolution image that Michael had downloaded from NASA. Intrinsic presented this demo at a trade show called SIGGRAPH, the annual gathering of the 3-D-visualization software community, in Los Angeles in 1999.

Intrinsic's CTFLY demo ended up having one problem: It was too good. Presentations to potential Intrinsic Graphics software customers inevitably veered into free-for-all geography classes, devolving into a tour of the globe, instead of a tour of Intrinsic software. Nevertheless Michael and Brian continued to invest in CTFLY, fine-tuning the demo despite the fact that it wasn't core to the company's game development software. After multiple quarters of working on CTFLY, the Intrinsic board of directors came to view it as an expensive distraction and gave the team an order: "It's cool, but stop working on it."

If CTFLY couldn't be worked on as a demo, could it work as its own company? Brian and Michael had witnessed the enthusiastic reception to the technology and couldn't let the concept die. Michael returned to the board and asked, "Can we try to spin the technology out to raise money and to license out the core technology to a new company?" The board agreed.

To run this new company, Brian and Michael knew that they

would need to hire a CEO to raise capital and build the team. They retained a Silicon Valley headhunter, who forwarded multiple interesting CEO candidates, including one who had recently sold a video game business he had started while in business school. One of Intrinsic's early employees, Andria Ruben, reviewed the candidate résumés and noticed that John had gone through the UC Berkeley Haas MBA program at the same time as her brother Ed, both graduating in 1996. After she had gotten her brother to vouch for him, John Hanke was scheduled for an interview with Brian and Michael in December of 1999.

Based on the introduction, John traveled from his home in the East Bay down to Mountain View to meet with Michael and Brian. At the first meeting John saw the CTFLY demo and heard the pitch from Michael about what he believed it could be. John pressed a bit deeper. "So I see the chip of Denver here. Do you have any other data?" Michael answered, "No, but that's no problem." John continued: "I see this is running local on a single machine. A really beefed-up one at that. You are saying that you think this can run on a *normal* consumer machine *over the Internet*?" Again, Michael answered, "Yes, that's solvable." In response, John asked, "You mentioned roads and showing other kinds of data. Do you have any of that working?" Again the answer was "Not yet, but we think that's solvable." John left the meeting impressed by the demo, but aware that a massive amount of work would be needed to turn the concept into a consumer product.

A few days later, John returned. "I'll do it," he said, "but there is something I want to tell you. If you ask me, you should shut down what you are doing and focus solely on this. I think it has much more potential than the game engine you are working on. And you two should run it. You don't need me."

Michael and Brian looked at each other. As much as they liked CTFLY, it wasn't the bet they wanted to make with their VC investment. Brian replied with a defense of the Intrinsic Graphics business model: "The game industry is a multibillion-dollar industry—and it's screwed. None of the platforms are compatible. Developers have to rebuild their games for each platform. They waste millions of dollars. We are going to give them a way to write the game once and have it work everywhere."

If John wanted to take on the "Earth" project, the message was clear: It was his to run and it was up to him to land the venture capital funding that would be necessary to turn the demo into a product and ultimately a business. In addition to the creation of the software, massive amounts of data would have to be obtained. Tools would have to be written to process it and servers built to host it. And then there was the matter of a business model to support all of this investment.

Later, Michael recalled the decision to hire John this way: "We knew there were going to be a lot of obstacles, and you could just tell John was a guy who would figure out a way of making it happen." During those interviews, he recognized John's perseverance and grit.

After John signed on, his first job was to recruit a team to turn the demo into something real. The only Intrinsic employee who would become part of the new company was an intense and fiery engineer named Avi Bar-Zeev, who was heading up development for the CTFLY client application. Avi was supremely talented, having just come off a project for Walt Disney Imagineering that utilized a high-end SGI machine to create a simulated 3-D river-rafting experience. Although talented, Avi was not getting along with the rest of the Intrinsic team—something that John would only learn about several months later.

The other first hires—Mark Aubin, Chikai Ohazama, and Phil Keslin—had all worked together at SGI. Redheaded, bearded Mark was a Silicon Valley engineer–meets–Northern California freethinker; he owned land in the Santa Cruz Mountains and gardened, and his children were homeschooled. He was a resourceful jack-of-all-trades among software engineers: He could build a server from its parts and write code, and he didn't mind putting together office cubicles, if that was what was needed. Mark was charged with processing what would eventually be the terabytes of data that would flow into the Keyhole database.

A young, hardworking Japanese American, Chikai graduated from Duke University with a PhD in biomedical engineering. There he had worked on 3-D visualizations of the human body—for example, taking the data of a heart and creating a 3-D model for medical study. Chikai was also a musician; interestingly enough, many of the best software engineers in Silicon Valley are also talented musicians. At Keyhole, Chikai created the tools to process data. This was a critical step in expanding the CTFLY demo from something that incorporated data from a single location to one that could include data from many places around the world and eventually the entire surface of the planet.

Phil was a software engineer who grew up in Dallas and graduated from the University of Texas with a degree in computer science. Phil had perhaps the biggest job of all. The problem with the CTFLY demo was that it was just that—it was only a demo.

All of the data was loaded onto the demo machine where it could be read quickly and directly from the computer's hard drive. The promise of Keyhole was that vast amounts of data could be collected and processed into a central database, hosted on a server, and then streamed to users over the Internet. This capability was just a theory; it was Phil's job to make it a reality. He would architect and build the system that turned Michael and Brian's demo and theories into a marketable service: a massive model of the planet that could be streamed over the Internet, utilizing special networking code. As a result, a consumer anywhere in the world with an Internet connection could access a vast and incredibly expensive trove of data and fly through it fluidly as if it were a local application.

John, Chikai, Phil, Mark, and Avi were the official founders of Keyhole, or the original five. Everyone else would come after John secured funding. They called their spinout project Keyhole, a nod to a secret system of U.S. spy satellites. By the late 1990s, the eleventh generation of Keyhole satellites (KH-11) was orbiting in space, dutifully capturing surveillance images of international hot spots. The Keyhole name was meant only as a placeholder until John could secure the venture funding and the earth.com domain could be acquired.

My role with Keyhole hinged on the venture funding. Throughout the fall of 1999 and the spring of 2000, I continued working as the marketing director at the *Statesman* website in Austin, and I told John that I would accept a job at Keyhole as soon as he closed on a round of funding. We agreed that he would include me in investor pitch decks as part of "The Team" as VP of marketing for Earth.com. From my perspective, since the financing had not yet closed, it was still too risky to move to California. Shelley and I had just gotten married and I didn't want to leave my job—or ask her to leave her job—without a financial safety net.

In the spring of 2000, on one trip out to California, I met John and the rest of the team at a building in downtown Oakland to look at an office space on the very top floor; everyone loved the idea of an Earth.com sign shining prominently on the city's skyline. Truthfully, the team was enthusiastic about moving anywhere and getting out of that crowded Intrinsic cube to start the new company in earnest.

But the NASDAQ was gyrating wildly, so start-up investors were running for the hills. To make matters more challenging, Keyhole was a brand-new concept: a revolutionary idea, yes, but also borderline experimental. The technology was an unknown in terms of its uses and market opportunities. More important, it could run on only the newest personal computers, those that had been built during the past six months. Potential investors were scared off when they were unable to get the software to work on their PCs.

As the spring of 2000 turned to summer, several investment leads evaporated. There was no earth.com domain or any fancy office space yet. The small Keyhole team continued working on the project within Intrinsic while waiting on answers from VCs. They refused to let the idea die, and John had promised to hire me as soon as he landed the investment deal. This became a cruel joke between Shelley and me—the investment and a new job on the West Coast were always three weeks away.

My commitment to Keyhole soon wavered.

A former boss from the *Statesman* contacted me one day about a marketing job in Boston. His company was one of the last to secure investment capital in the fall of 1999, locking in a whopping $74 million investment from Charles River Ventures to start yet another Internet marketing consulting firm. Shelley and I flew to Boston in June, where I interviewed, caught a Red Sox game, met the founders of the company, and was offered a job while waiting for my flight at the Dunkin' Donuts in Logan Airport. The number presented to me on a folded-over Dunkin' Donuts napkin was a lot more money than I was making, so I took the job and would start in July.

The call to John was hard: I was certainly more excited about the Keyhole opportunity than the Boston job because I knew that the Keyhole technology had a chance to become something transformative. John was disappointed but understood; and, truthfully, the Keyhole software product wasn't anywhere near ready to be "marketed" to anyone. I wished him well. We agreed to keep in touch, and I offered to continue to assist pro bono on any marketing projects for Keyhole.

While I was ramping up on my new job in Boston, John was finally making headway on the fundraising front. In late 2000, Sony Venture Capital had committed to do the Series A investment, but another swerve in the markets caused a delay. It was scheduled to close in January of 2001.

In December of 2000, John and I met up at the Holiday Bowl in San Diego to watch the Longhorns play Oregon. He was relieved to finally have funding commitment. The Keyhole team would soon be officially on its own and could pay Intrinsic back for its seed support. As we walked around the city's Old Town neighborhood, I was negotiating for cheap tickets to the game and John was negotiating with his new landlord. The landlord had seen the Keyhole demo and wanted some shares in the company in exchange for the lower rent that he claimed to be charging. In front of a lively Mexican taco bar, John gave me a thumbs-up without breaking stride in his conversation. We sat down at a table and I ordered two Budweisers. The call continued through our first beer. Hanging up the phone with his landlord, John sighed, shook his head, grabbed a beer that I slid in his direction, clinked mine, and said, "Done!" Taco after taco, beer after beer, we caught up.

Listening to John that day in San Diego, I was relieved to have opted for a steady paycheck from the consulting firm in Boston. It had been a long six months for John and the Keyhole team. In addition to the lack of capital, personality conflicts with Avi strained the team. To Avi, John had said more than once, "You're going to be able to do more if you learn to work with people versus working on your own." Things were at a breaking point with Avi, but despite this volatile work environment, John had learned to manage his star software engineers: how to recruit them, challenge them, and keep them. Tensions ran high, but somehow the Keyhole service was becoming more of a reality.

Building on the early concepts of Intrinsic, Phil, Avi, Chikai, and Mark had solved a problem that would forever change the way maps were used over the Internet. It started with preprocessing vast amounts of data on a server to create an optimized "quilt" of mapping data tiles at multiple resolutions blanketing the entire Earth. Then the team built a sophisticated client software application users installed on their computers that could retrieve and render that data fluidly. This type of software architecture is called a thick client; you are accessing something living on a server, but your client application on your computer is doing complex computational heavy lifting, too. With a reasonable Internet connection, users could now fly through a map of virtually unlimited size without any delays to stop and load new data sets. A computer of modest capability could appear to have the power of a supercomputer; mapping data moved cleverly from servers at just the moment it was needed.

By the end of 2000, John had even begun to send me actual software bits—executables to install and run. Unfortunately, I didn't have one of the 15 percent of the world's computers that were powerful enough to run the software.

Somewhere around the third beer, I asked John: "Why don't you try to build a web-based version of this thing so that *everyone* can use it? I know that it wouldn't be the same fast and fluid 3-D experience as the downloaded software application, but visiting a URL is a whole lot easier than getting someone to install software."

As football fans loudly cheered their teams on and a mariachi band played not far from our table, John explained why a web-based version of Keyhole was a risky strategy. It was a very expensive turn down a street with lots of competitors. "Our differentiator is the fluid 3-D animation," he said. "That's what makes it magic. We can't do that in a web browser. If we put it in a browser, it will be just like MapQuest, and they already own that market."

MapQuest—which was bought by AOL in 1996 for slightly less than $1 billion—had become a verb for mapping before Google even existed. Because MapQuest represented 90 percent of the market share in the United States, "I'll MapQuest it" was a common refrain for someone who was looking to find his or her way.

"I don't want us to build a product for today's computers. We want to build a product for where technology is going," John explained. At the time, the Keyhole board included Brian McClendon as well as the Sony investment representatives. (Intrinsic got a board seat for incubating the idea and contributing the initial intellectual property.) He told me about new 3-D graphics cards in mainstream PCs from companies like Nvidia and ATI. He told me about fast broadband Internet access, and about more powerful mobile devices and far faster wireless networks. He leaned forward, his voice rising above the mariachi band.

"You know these things have GPS chips in them now," he explained excitedly, holding up his Motorola flip phone. "It's required by law, and all the handset makers and wireless carriers are going to have to comply with the requirement that 911 calls from mobile phones have to be able to be located. Can you imagine what that is going to mean?"

Uh, nope.

John was predicting a future that I couldn't see. Today we take those powerful computers, that broadband Internet, the omnipresent iPhone for granted. In 2000, very few people owned a computer powerful enough and had Internet access fast enough to even get Keyhole to run. I took another swig of beer.

I had seen, and even contributed some marketing hoogity-poogity to, the early Keyhole investor pitch decks. The slides had included heavily Photoshopped images of Keyhole EarthViewer running on desktop computers ("okay"), notebook computers ("well, not mine, but okay"), *and* even, most ridiculous of all, mobile devices ("you've obviously been watching too much *Star Trek*").

Later that night, after the Holiday Bowl and the last-second 35–28 Longhorn loss, Shelley and I talked. "Yeah, I don't know where he's going with this thing," I whispered to her on the phone because John was just out of earshot in the hotel room. "I'm glad that I have my job back in Boston."

Chapter 3

THE GREEN FOLDER

"You are totally safe," my boss in Boston assured me.

It was late March of 2001, the end of the quarter, and the Internet marketing consulting firm where I was working had not made its projected numbers. Not even close. The burst of the Internet bubble was official, and the director of human resources had been spotted numerous times at the copy machine, busily copying and organizing severance package documents into tidy green folders.

My boss had told me that my job was important to the company, which meant the call to the CEO's office was unexpected. I was out. I walked to my boss's desk with my green folder and said, "Hey, WTF? I thought you said I was going to be safe."

"I know," he said. "I thought I was going to be safe, too." He held up his green folder and smiled.

It had been one of the coldest winters on record. Even though it was March, the city was still covered in dirty snow and ice. It would be a few days before I got the gumption to call John to tell him the news. "I was voted off the island," I joked, trying to make light of the situation.

While the recent Series A investment from Sony was short of the $10 million the company thought it could raise during the height of the dot-com frenzy, John did now have a bit of working capital. The timing of the investment turned out to be a minor miracle for the company—and for me!

"Maybe you could do a bit of consulting work remotely for us and we could see how it goes," John offered. We both agreed full-time would be challenging if I remained in Boston. That said, I was happy to join the eleven-person Keyhole team, even on a contract basis.

Unfortunately, just as I joined, Keyhole's business plan had been turned on its head. John and team had spent the first year in business creating a broadband mapping service for consumers, and had a deal in place to distribute the soon-to-be-released Keyhole EarthViewer through the leading broadband Internet company at the time: Excite@Home.

Excite@Home had millions of broadband Internet subscribers, and EarthViewer appeared to be a perfect fit for these tech early adopters who were paying for faster Internet service. Excite@Home's business, however, was deeply dependent on the cooperation of cable companies like Time Warner, Cox, Comcast, and others. As the dot-com craze came to a fiery and final end in early 2001, the cable companies pulled out. Excite@Home's stock cratered, and, like many others, the company began splintering apart. Keyhole had hitched its marketing-and-distribution wagon to the wrong horse. It was easy to do. There were a *lot* of wrong horses in early 2001. In fact, just about any dot-com consumer Internet service was suddenly a risky bet. From Garden.com to Pets.com to iWon.com to any of the other sixteen dot-coms buying Superbowl ads in January of 2000, over $5 trillion of market valuation evaporated by the end of the year.

John and the Keyhole team needed another business model.

A new idea other than consumer mapping software. "Hey, we need a new business model," John said to me on our kickoff phone call. I stood in the food pantry that was now doubling as my office in our Cambridge apartment.

"I've been researching GIS software companies," I said a few days later on the phone. (GIS stands for geographic information system. It is *enterprise* mapping software used to create and analyze data using maps.)

"Yeah, I know a little bit about it," John said, "but I don't want us to get pigeonholed into trying to sell something to the government. Plus, the market is owned by one company."

John was referring to Environmental Systems Research Institute, or Esri (pronounced *ez-ree*), the leading mapping software company. What I didn't know at the time was that the entire digital mapping software industry had gotten its start thirty years earlier about half a mile from where I was sitting: at the Harvard Graduate School of Design, with a landscape architecture student named Jack Dangermond.

Dangermond and his wife, Laura, created the software while at Harvard, then used it for their land-use planning consultancy in the 1970s. The husband-and-wife team landed a huge project for the county of San Diego in the early 1980s, and their work morphed the tool into a digital mapping software product that they called ArcGIS. By 2000, Esri had hundreds of thousands of clients that produced over a million maps. Daily.

In addition to cranking out maps, the company cranked out cash: In 2001, Esri generated $300 million in revenue; Jack Dangermond, who still owned 100 percent of the company, had a net worth that Forbes estimated at $2.7 billion.

Esri had a mapping solution for every industry. Police departments used it to map crime incidents and pull in feeds from the FBI and other government databases. The military employed the software for change detection, allowing an analyst to compare two different satellite images of the same place and detect what had moved (for example, a tank or a missile). Tax assessors utilized it to examine comparable home values and plat maps outlining a property. Real estate brokers created heat maps representing the concentration of a certain data set, and used drawing tools and regression analyses to determine the optimal location to build the next Starbucks or Home Depot. The applications were endless.

Add that to the fact that Esri had an entrenched installation base of system integrators, sales reps, and long-term contracts and service agreements—and it was a daunting idea that we might try to crack into the GIS software industry with our incomplete alpha version of Keyhole EarthViewer. No wonder John was dubious.

Nonetheless, I continued to push the GIS market because I thought we might have a competitive advantage. Esri, for all its merits, was a traditional enterprise software model and had three shortcomings: It was complicated, had no data, and was slow.

Creating maps using Esri software required sophisticated data analysis and training. You could get an accredited college degree in GIS, spending four years learning how to use it. More realistically, a team of Esri-trained specialists (or Esri's own consulting arm) was needed to configure a custom solution for a client.

Esri was essentially a blank sheet of paper. In order to begin using it, you had to go out and find the data you needed, and download and import it. In the meantime, you had to hope that the data was the right format and map projection so that the multiple layers of data worked together. Esri specialists were needed to acquire and integrate mapping data in order to make the application work.

Lastly, it was slow. Viewing an Esri-generated map online involved a click followed by a long, painful wait for the map to refresh. If you wanted the base map for your Esri map to be an aerial image, the software bogged down even more. Esri was glacially slow, almost unusable, especially if you didn't have the aerial image loaded on your local hard drive.

By comparison, Keyhole was simple to use, came bundled with access to terabytes of data, and was lightning fast.

In the summer of 2001 during one of my monthly trips, I visited a real estate company in San Jose to learn about how they used GIS software. The mapmaking team included only two specialists trained in the use of Esri software. As the pair escorted me to their rear office, I walked past the sprawling cubicles of one hundred and forty real estate brokers. During our meeting, several of the brokers interrupted our conversation with urgent requests for specific maps.

Wait a second, I thought to myself. *What if we could make a GIS mapping application that was simple enough to be used by these brokers?*

At the Keyhole office later than day, I recounted my experience for John: "Maybe Keyhole could be a simpler GIS aimed at non-GIS experts?" I joked, for effect. "Maybe we could be the GIS solution for people who can't spell GIS."

It wasn't as if John was unaware of this potential market.

Others on the team had also been advocating for an enterprise GIS strategy. Although the long-term potential was more limited, the promise of an immediate cash flow was increasingly attractive. With the collapse of the dot-com bubble, future funding was looking less likely and the Keyhole bank account was shrinking every day as the company invested in servers, data, and more engineers.

Since the finalization of the Sony investment deal, John had begun to fill out the team.

A software engineer named David Kornmann, who had worked with Rémi in France, was hired (although he continued to live in France). David had seen the Intrinsic CTFLY demo at the SIGGRAPH trade show in Los Angeles in 1999 and was duly impressed. Rémi had given David a copy of CTFLY on CD-ROM to take back to France with a challenge: Could David get 3-D terrain running in the software? Rémi was well aware of David's love for 3-D terrain visualization projects. They had worked together on an Airbus A320 flight simulator, and David had shown his passion and expertise in the complex software code behind the accurately visualizing topography, making the mountains pop off the surface and the valleys plunge below.

Back in France, David added small examples of 3-D terrain code to the demo and sent Rémi a new executable program on CD-ROM. CTFLY had always allowed the user to zoom in to a place on the earth, which usually was the first level of amazement. Months later, David would come through with a demo of a 3-D terrain rendering of the Grand Canyon and Mount St. Helens for Michael and Brian. Though he still had much work to do to make it something that could be utilized worldwide, in

David's demo the user could zoom in to a place on the earth and *tilt* the view to render terrain in 3-D, offering a whole other level of realism to the scene. Mount St. Helens snapped up from the surface, the Grand Canyon pitched downward.

At the same time, John hired his first administrative assistant, Dede Kettman, a tall blond Italian woman who was always dressed in refined business attire. Visitors were treated with an air of decorum and hospitality usually reserved for formal corporate meetings. In some ways, she ended up serving as the mother figure of the office; it was quite possible that Dede was too good for Keyhole. Lenette Posada Howard, a veteran tech project manager with a boisterous sense of humor, was hired as head of operations. Lenette had worked for ten years for groups like Anderson Consulting, project-managing a variety of software implementations, but agreed to work for Keyhole part-time as long as she could bring her two-month-old daughter, Gaby, when she worked in the office; she often had an eleven-by-seventeen Microsoft Project schedule in one hand and her daughter in the other. When the team was on schedule, which was never, Lenette's job was easy and everyone was happy with everyone else.

After I was hired, John also brought on Dave Lorenzini, a smooth, excitable, geospatial business development and sales representative who had jumped ship from one of our aerial imagery providers. When his combination of industry knowledge, contacts, and high energy was well directed, doors opened up. We rarely knew what Dave was doing, but we knew we'd hear from him once a week with some crazy new idea or opportunity. Living partly in Los Angeles and Lake Tahoe, and on the road, Dave was everywhere except where you expected him to be.

John also recognized the strategic importance of data acquisition and hired a handsome and somewhat regal South African named Daniel Lederman to lead overall business development. He was one of John's most critical hires. Think of it this way: Keyhole EarthViewer without bundled data was like iTunes without any music, a Kindle app with no books, or a YouTube player with no video. For maps, the data is everything. Daniel scoured the globe for free and paid sources of aerial imagery and other data sets in order to build out Keyhole's library.

Ed Ruben was a jolly, good-natured family man—probably even before he had a family. He earned a master's in computer science at UC-Davis and attended business school with John at Berkeley before working on databases for Netscape until John hired him as a Keyhole engineer, where he would develop our subscription billing system, a relatively new concept in the software world.

Soon, Daniel started acquiring data sets by the megabytes (individual aerial photographs), gigabytes (neighborhoods), and terabytes (cities and states). Many data sets were captured with taxpayer funds and therefore a part of the public domain, and Daniel could request specific sets of data—and have them sent to our offices via a CD or DVD for only the cost of reproduction. Surprisingly, in many cities, departments, such as police and fire, sent us their city's imagery data and asked us to import it. This allowed them to access the data themselves through Keyhole EarthViewer, something that they would have gone through the GIS department for before.

The entire base map for Keyhole EarthViewer 1.0 was free, courtesy of NASA and a data set called Blue Marble. It was a compilation of satellite images gathered from June to September 2000. In EarthViewer, you saw this image when you zoomed all the way out into space. The Blue Marble data set was a beautiful base map and freely available to the public. Of course, this was just the background image; it was not high resolution. For higher resolution views, we needed to find imagery captured by advanced high-resolution imaging satellites or low-flying planes.

John developed a three-pronged data-acquisition strategy. The first tactic was related to satellites. John and David Lorenzini reached out to the only two companies operating high-resolution imaging satellites—Space Imaging and Digital Globe. Both companies operated satellites that were the direct descendants of technology pioneered for the military, or the Keyhole satellites.

Only ten years earlier, Congress had spawned the development of a commercial satellite industry, passing the Land Remote Sensing Policy Act of 1992. For the first time, non-defense-industry, high-resolution imaging satellites were legal.

David and John met representatives of Space Imaging, a joint venture company formed by Lockheed Martin and Raytheon that had launched the first commercial high-resolution imaging satellite. Known as Ikonos, it could record images at a resolution of approximately one meter by one meter, meaning each pixel in the image corresponded to a spot on the ground about one square meter in size.

Digital Global was a long shot. The first two launch attempts had been expensive failures at $500 million per satellite (it *is* rocket science, after all). Fortuitously for "DG," as we called it, and for Keyhole, its third satellite named QuickBird II launched from Vandenberg Air Force Base on October 18, 2001, and successfully transmitted its first satellite imagery a few days later.

Designed to orbit five years, QuickBird II ended up flying for thirteen years, circling the Earth over seventy thousand times, continuously snapping photos at 0.7-meter resolution: This meant that one pixel in the photo represented 0.7 meter on the ground, higher resolution than Space Imaging's Ikonos. QuickBird's images were good enough to distinguish the difference between a car and a truck and the vehicle's color, though not exactly clear enough to pick out the make and model.

A second element of the data acquisition strategy centered on imagery collected from aircrafts for local governments. Aerial imagery could be even higher resolution than satellite imagery, with photos up to fifteen centimeters by fifteen centimeters per pixel, and much of this data had been commissioned and paid for by local government agencies. Keyhole began an aggressive business outreach program to get these organizations to share that data. In exchange, the cities were granted licenses to their own data through Keyhole's revolutionary streaming software platform.

The third component of the strategy involved going directly to the companies collecting aerial imagery and negotiating to purchase it. At this point, the industry was highly fragmented and was dominated by modest outfits with roots in aviation. One of the most colorful of these aerial imaging companies was a small shop with big ambitions run by the incomparable aerial imagery cowboy and risk-taking entrepreneur J. R. Robertson, CEO of Airphoto USA. Based in Phoenix, he maintained a fleet of fourteen planes with cameras mounted through holes drilled into the fuselages. Long-haired, chain-smoking, hard-drinking, Harley-riding J.R. was the antithesis of John and Daniel. But it required a renegade like J.R. to take a chance on Keyhole. He owned high-resolution aerial imagery data sets of over a hundred of the most populous cities in the country—and John wanted access to it all. He also wanted J.R. to deliver new city data sets to us within four weeks of flying. He wanted Keyhole to have the right to import all of Airphoto's imagery into our databases and the right to sell access to it. J.R. might charge a city government a fee of $400,000 to fly over a single urban area. And John wanted all of this for no cash up-front.

Instead, John would pay J.R. royalties: 25 percent of every license sold. For example, if we sold a $600 license to Keyhole EarthViewer, J.R. earned $150. J.R. probably agreed only because he could see that it was a faster and more efficient way of distributing his data, and if he didn't do the deal with Keyhole, someone else would.

Since we would be selling this service to many of his own commercial customers, J.R. negotiated to make Keyhole supplemental, not competitive, to his core GIS user market. For example, the contract stipulated that we would not allow users to export an image with geodata (the latitude and longitude) from Keyhole; this meant a mapping software application, like Esri, would be blind to the original location of the image. The printing resolution would also be arbitrarily capped, and, in some versions, watermarked with Airphoto and Keyhole logos. Lastly, the Airphoto logo would be displayed prominently anytime his imagery was being viewed.

Mark Aubin, who sat in a cubicle stuffed into the back corner of the office, transformed J.R.'s images and the images from government agencies into EarthViewer with his new tool called Earthfusion. It included back-end scripts for importing, stitching the photos together with the right map projections, color-balancing the images, and blending it all into a database in a format that could be streamed to EarthViewer client applications. As a result, it looked like you were viewing one seamless photo even though it was made up of hundreds of thousands of photos.

Keyhole EarthViewer was effectively two products: It was a software package that you downloaded and installed on your computer, and it was a subscription service to a library of aerial imagery. (We were intentionally staying away from building a website, so we wouldn't be in direct competition with MapQuest and also for security purposes, because the imagery data could be easily stolen from a browser-based service.) In addition, the pricing model was revolutionary. Up to that point in time, a single satellite image of an area just eight kilometers by eight kilometers could cost $10,000 or more.

We pivoted toward a business-to-business model, and Keyhole began to explore potential enterprise markets. Dave Lorenzini had established a connection at a trade show, Realcomm, known as a technology showcase for the real estate industry. That year it was hosted in Dallas, Texas, on June 14, 2001. John stood at the doorway of his dark gray cube in late May and told a small group of us the following: The Realcomm trade show would be the site for the official launch of Keyhole EarthViewer 1.0.

Mark said, "We're not ready for that. We have only four cities."

"At least have Dallas/Fort Worth uploaded in the database

before the show," John said. "Look, guys, you have to figure out a way to get this done." Clearly, it was not up for debate. The engineers worked late nights to ensure a new database was ready for the show, and concurrently another team finalized the EarthViewer 1.0 software. I scrambled to prepare all of the necessary marketing elements—from business cards to the trade-show booth. In terms of price, we decided to try an annual subscription of $1,200.

On the day of the show, two thousand investors, developers, and real estate portfolio managers assembled in the enormous hall in Dallas to hear the latest industry technology insights from the show's charismatic organizer, Jim Young. On a stage flanked by two video screens, Young ticked through the most recent innovations before introducing his latest tech discovery—Keyhole EarthViewer. Chikai began a flyover of Dallas and zoomed down to the convention center. A tour of real estate hot spots across the country followed with Young expounding on the potential of Keyhole to revolutionize the way real estate was discovered, acquired, and marketed.

Later, during the trade show, I was in the Keyhold booth, flying a potential client around the globe and into aerials of his properties. I sensed J.R.'s presence nearby. I knew what was coming before it happened. The Airphoto USA logo had disappeared just as I piloted the prospect down to a property.

"Where's my logo?" J.R. whispered into my ear from behind me. "Where's my fucking logo?"

Then he stomped off in his cowboy boots, his wallet chained to his jeans. Soon after, I received a call from John, asking me what was going on. Back in Mountain View, J.R. had left him a menacing voice mail about "breach of contract" and "a clear violation of trust." Later, I found J.R. in his booth, sipping whiskey from a plastic cup and smoking a cigarette.

I tried to explain the reason for the disappearing logo—the disparate data sets and how sometimes our client software didn't call back to our servers for a new image. In those rare cases, his logo might not load because EarthViewer did not know that the image being viewed was from Airphoto USA. We only showed his logo when looking at one of his photos.

"Desperate databases?" he asked.

"No, disparate," I said, "as in different."

"Why don't you just fucking say *different*?"

Despite the glowing reviews from Realcomm, the team returned from the show with only eleven sales. John tacked this list of subscribers to the wall outside his office. We did receive a show award for Best New Real Estate Technology, but it was clear that we had much more work to do to find buyers. We managed to make more money than we had spent, and this became the standard for how we decided on whether or not to participate in a trade show. An architecture show? The cost $5,000? "Okay, you better come back with more than that in sales, Kilday," John might say.

It may have been naive, or possibly brilliant, but we wound up trying out many different markets by simply going to the trade show of that market. Travel agent show. United States Geological Survey show. Television broadcaster show. Private aviation show. Energy sector show. Urban planning show. Military show. We weren't picky. It seemed that the technology could be used for a number of scenarios we couldn't have imagined. I remember selling an EarthViewer license to someone who designed highway billboards. He could use the tool to figure out the sight lines of a billboard and adjust the font size in his designs accordingly.

That fall of 2001, we flew all over the country to trade shows and sales meetings, trying to find the right market for our product. I was still living in Boston, as the company's financial prospects were too shaky to justify a cross-country move for Shelley and me. On Monday, September 10, 2001, I took my normal American Airlines flight out of Boston's Logan to California at eight o'clock in the morning. I remember that David Lorenzini had called me the prior week, asking if I could instead fly on the Tuesday, September 11, for a trade show that he was going to in Los Angeles. I had turned him down, though, as I already had made my arrangements to travel on the Monday.

What happened the next day put into perspective the relatively trivial trials and tribulations of life at a risky start-up. Like the rest of Silicon Valley and the world, we huddled around televisions and the CNN website. We were searching for answers: Who were these people? Where did this hate come from? CNN's website was down, overloaded with traffic. For the first time that morning, I spun the EarthViewer globe to study the Middle East.

Somehow I felt more committed to Keyhole after the tragic events of September 11. Like many, for me, the events and enormous loss of life on that morning served as a stark reminder that we only live once. Even though it was a risky venture, we had a good chance of turning Keyhole into something great. I felt more willing to take that chance.

Despite requiring users to own a newer PC in order to run

EarthViewer and having only a modest imagery database, by the end of 2001, Keyhole had generated $500,000 in revenue. The transition from the original strategy of a consumer-focused software company had been challenging, but we were reinventing ourselves in order to keep the doors open. Many dot-com consumer companies that had not pivoted to a new strategy were closing up shop all around us.

At the beginning of 2002, we hired our first dedicated sales director, Doug Snow. He had sold mapping software successfully before for MapInfo, the distant number two in the GIS software world. Doug, a former college football linebacker and Tony Soprano look-alike, brazenly offered to work for Keyhole for commission only. I also managed to convince John to hire a marketing coordinator, Ritee Rouf.

That spring I began to get the sense that we might gain greater traction in real estate. That said, real estate is a vast and sprawling industry; it's really a loose collection of many different markets, each with their own needs. First, we tried *residential* real estate with the National Association of Realtors show in Chicago that attracted fifty thousand residential Realtors. Even as I was unpacking our booth at the back of the convention center with Ritee, I knew the residential real estate market was not going to work out. Being located next to a lipstick-sales company was my first clue.

In hindsight, I caught a glimpse of the future during those painful two and a half days at the National Association of Realtors show. The mostly female crowd wasn't buying Keyhole. They *were,* however, carrying two devices: their ubiquitous mobile phones and also new mobile personal digital assistants

(PDAs) from Palm Pilot or Compaq. Some of those PDAs even had the ability to pull up a tiny black-and-white map, if you were patient enough.

A month later, we tried our luck at the largest *commercial* real estate trade show in Las Vegas: the International Council of Shopping Centers (ICSC) show attended by fifty thousand commercial real estate brokers. Our booth was located on the outskirts of the convention center across from a company that sold LED road signage, the kinds of signs strip malls and gas stations display so you can see them from great distances. Like space. One of their employees even offered us pairs of sunglasses due to the bright lights of their booth. Within five minutes, I felt like I was getting a tan.

"This is total bullshit," Doug Snow exclaimed and stormed off. Fifteen minutes later, we were escorted to an optimal spot next to the front door of the exhibition hall. It was the single most important thing that Doug did for Keyhole. The company was given that prime location—as well as the first right of refusal—every year.

ICSC 2002 proved to be a coming-out party of sorts for Keyhole. It was the first time that people bought our product in volume. It was a feeding frenzy of sales. In three days we'd sold $100,000 worth of recurring software subscriptions and generated lots of leads.

"Show special $599. Come on, let's do this. Hell, you are going to spend that much money on a steak dinner tonight," I teased a prospect. He bought the software. (We had dropped our price from $1,200.)

It was almost always the same. "Oh my gawd. Can I see my

house? Jimmy, get over here and check this shit out. Hey, that ain't your car, Jimmy. Hell, you better call your wife. Hey, this ain't real time, is it?"

"Sure it is," I responded. "Go on outside and wave up at the satellite. We'll stay in here and watch you!" He looked at the door and then back at me.

"Awww, hell, you almost got me, Billy! Okay, sign me up. I'll take two."

In those days we used an old-school credit card slider that made carbon copies. Dede took those credit card orders and entered them into a website to process the purchase. Our booth was flooded with sales to the point that our staff was fighting for the credit card sliders. We had a contest at the booth about who could sell the most each day, and Dede often won.

I once held up four software packages above the crowd and said, "I've got four packages left. Who wants them?"

"I'll take them," someone shouted from the back. As I rung the customer up and handed over the software, he said, "Hey, what is this, anyway?"

Keyhole had found its first market.

Chapter 4

OUT OF GAS

On a brilliant Saturday morning in the late spring of 2002, I found myself stuck inside the air-conditioned atrium of the Tech Museum of Innovation in downtown San Jose. Before the doors opened, I reached into my backpack and dragged my Toshiba Satellite laptop computer out, setting it up on the standard six-foot, black-draped table.

I would have preferred to have been anywhere else, but especially back in Boston with Shelley, who at that point was seven months pregnant. The Tech Museum in San Jose had recently signed on as a customer, installing Keyhole EarthViewer in a kiosk to allow individuals to explore the planet. Since I was in California on the tail end of one of my monthly treks to Mountain View, I had agreed to demo it as part of the rollout.

I was tired. The commutes to Mountain View were intense, seemingly working in three weeks of action into one weeklong sprint. Without Shelley around to curb my long hours, I was free to work unabated, late into the night. I often slept on the pullout sofa in Holly and John's dark basement in Oakland, driving the hour and a half each way with John in his silver Subaru WRX in order to save the company money on hotels; the commutes were mere extensions of the workday, filled with sales calls and strategy planning as we slogged down I-880 in bumper-to-bumper traffic and across the Dumbarton Bridge into Mountain View, the heart of Silicon Valley. Occasionally that summer, John and I escaped to catch a game played by the Oakland A's, who were inexplicably in the midst of a twenty-game winning streak, later immortalized by Michael Lewis in his bestselling *Moneyball.*

By Saturday morning, I was fried. I fumbled with the VGA cables to connect my laptop to the fifty-inch display monitor. Soon Keyhole EarthViewer floated on the screen. I gave the globe a spin with a casual flick of my mouse. I looked up across the atrium and smiled knowingly at the janitor tidying up before the doors opened. He paused his broom mid-sweep, and his eyes grew wide.

It never got old: demoing Keyhole EarthViewer. The reaction was often religious. The doors opened, and the atrium bustled with Silicon Valley's above-average kids and their above-average parents chasing after them. The game was on. Audible gasps. Slaps at my shoulder in disbelief. Expletives, from the children. Offers to invest, from the parents. The sheer joy and the jolt of adrenaline always re-energized me.

Later that morning, John sneaked up behind me with his five-year-old son, Evan. (At the time, Holly and John also had a two-year-old daughter named Claire.) "You know, it's rare to get to show something to someone that they have never seen before," he said. "Like, *never* seen before." He jumped in and took the controls for a few spins to show the gathering crowd some of his favorite sights, such as the Grand Canyon and the strip of Las Vegas.

The crowd surrounded us. The feeling of euphoria was palpable and infectious. "We've got to get the consumer version out there," John added, looking around at the awestruck crowd clamoring for a chance to control EarthViewer.

The commercial real estate market had been a boon for Keyhole, no doubt. They were the first guys to step up and write our fledgling company checks, providing us with invaluable cash flow and extending our runway a bit farther. But facilitating a land deal for Home Depot or helping an HVAC installer figure out the pitch of a roof or assisting Starbucks in choosing their next location was not exactly the vision that got John out of bed in the morning.

Yes, those first few thousand customers were critically important, but as John surveyed the families in the museum atrium that day, he was drawn back to the original idea: a fully immersive, fast, fluid 3-D model of our common planet, shared by everyone. The revolutionary concept that anyone, anywhere could virtually fly across the planet, using a model so realistic, it would be as if you were there.

John looked around the enthusiastic crowd again. "Yeah, we've *got* to get the consumer version out," he repeated, this time a bit impatiently. The desire to create a consumer version wasn't just about pleasing the crowd; Keyhole desperately needed the cash.

John had been recently introduced to Jen-Hsun Huang, cofounder of a company called Nvidia, the leading manufacturer of 3-D graphics processors. Michael Jones had connected with Jen-Hsun in January of 2002 at the Consumer Electronics Show (CES) in Las Vegas and raised the idea of Nvidia partnering with Keyhole to promote its graphics technology. The seed was planted, but it was up to John to take that introduction and transform it into a deal.

Huang started Nvidia in 1993, just as the PC-gaming industry took off and those same Intel-based PCs that put SGI out of business hit the market. The company popularized the graphics processing unit (GPU), a computer card dedicated to crunching the complex math behind the creation of three-dimensional gaming worlds for games like Doom and Quake on PCs from Dell and Hewlett-Packard. The GPU enabled intricate computer 3-D worlds to be graphically rendered on cheap, consumer-grade PCs, opening up whole new markets and industries. By 2002, the company was worth over $10 billion.

For all of its success, Nvidia appeared to have hit a ceiling on its market valuation. Even though it dominated gaming, which was a large and growing market, Wall Street still did not consider the company to be operating in a mass consumer market. For example, you didn't need a Nvidia graphic processor to run a web browser, open a spreadsheet, or read an email. Jen-Hsun saw Keyhole as a chance to move beyond the gaming market. EarthViewer was not a game, yet because of the complex 3-D math at its core, it did need a computer with a dedicated graphics processor to run.

During the initial meeting, Jen-Hsun had a question for John: Would Keyhole consider building a version of EarthViewer that was intended for consumers—one that was exclusive to Nvidia, *requiring* a graphics card made by Nvidia to run?

After conferring with Phil about the ease of technical implementation, John made the offer to Nvidia's head of business development, Jeff Herbst: Keyhole would give Nvidia an advantage over other graphics card makers by launching the consumer version of Keyhole exclusively with Nvidia in exchange for a million-dollar investment and a commitment to bundle the app with all graphics cards and software updates. "No way," the Nvidia team responded dismissively. "Don't you understand the kind of volume such exposure would drive? We're going to open up a firehose for you guys."

Keyhole was running on fumes, and John was relentless. We urgently needed the cash; the Sony funds were gone (and that venture capital fund was being folded into another department at Sony). He was confident that a version of EarthViewer exclusive to Nvidia would be a valuable differentiator for them, so he decided to reach out to Jen-Hsun directly. Eventually Jen-Hsun trumped his business development team and agreed to pay Keyhole $500,000 to build a consumer version "optimized for Nvidia GPUs." John was satisfied with this agreement because it meant another critical two to three months for Keyhole. Within weeks, Phil and our team created this version of EarthViewer, but devoid of some professional features, like printing, annotation, and measurement. EarthViewer NV, as we called it, was bundled with a fourteen-day free trial and a one-year subscription price of $79.95. Our aerial imagery partner J. R. Robertson received 25 percent, or roughly twenty dollars, for every sale.

To kick off the partnership, John and I met a team from Nvidia for lunch near their offices in Santa Clara. I was introduced to my counterpart on the Nvidia marketing team, Keith Galocy, a good-natured midwestern guy who was often bleary-eyed from all-night gaming sessions "testing out" the latest graphics cards.

After lunch we headed back to Nvidia's sparkling new offices to review launch plans. The enormous glass space was bright and futuristic. As a part of the move into these new offices, one of the Nvidia engineers had created a realistic first-person 3-D shooter game out of their offices, so employees could roam the hallways and blow away their virtual colleagues. (This seemed to me to be a terrifically bad idea.) Jen-Hsun himself dropped by the meeting to congratulate our team on the deal. Born in Taiwan and a graduate of Oregon State and Stanford, Jen-Hsun founded Nvidia on his thirtieth birthday. "Have you given any thought to 'going procedural' when you zoom all the way into street level?" Jen-Hsun asked, leaning forward over the conference table toward John.

"Possibly," John responded, "but we are committed to a geo-specific approach. We need things to look like what's actually there, and the data isn't out there yet to do that at the street level on a global scale."

On the five-minute drive to Keyhole, John explained to me what Jen-Hsun meant: He was predicting a world of 3-D buildings and street-level views. Jen-Hsun was proposing to do that procedurally, a computer graphics technique where you procedurally create detailed images from an algorithm. This could yield buildings that would make a vivid 3-D demo, but it wouldn't reflect what the real world looked like at that location (think Sim City). John said that it wasn't out of the question that one day a user would be able to zoom all the way in to a map to street level, and from there the user could switch over to a realistic view of 3-D buildings and street-level photos and be able to virtually walk down the street.

"This is assuming that you can get the millions of miles of street-level imagery data, of course," John said, and then after a long pause added, "It's a twenty-year-out project."

The idea sounded like something out of a science fiction novel to me. Certainly far-fetched technology that I would never see during my lifetime.

The consumer version for Nvidia was an easy sell. Consumers loved it. The following is an email sent by a college student after he had downloaded the free fourteen-day trial of Keyhole EarthViewer NV from the Nvidia website. The subject line read "Keyhole is Awesome." And here's some of his email: "OMG. I was on the Nvidia site, looking for drivers. . . . And I got to the screen and I saw the earth. It was floating there. OMG. OMG. I started drooling. . . . I zoomed, and Jesus, it was still high res so I went to my house, I went to my school, I went to my friend's house and then I discovered the address capabilities!!! And then all of these people came into my room and we typed in addresses and we freaked and tripped out. This program is crazy insane. Are you sure this is legal?"

The Nvidia deal would provide a helpful cash injection, but John knew that it was not enough. The $500,000 could only be stretched over two or three months. So he pursued another $500,000 deal with a company in Tokyo, Japan, called Silicon Studio, a video game developer that specialized in creating and distributing 3-D games. In early 2002, the company came to John with a desire to create a version of EarthViewer for their exclusive distribution in Japan. After weeks of negotiation, John had flown to Tokyo with the hopes of ironing out the details, finalizing the deal, and returning home with $500,000 more for Keyhole.

While in Japan, John received the heartbreaking news that his father, Joe, had fallen gravely ill back in Cross Plains, Texas.

As a result, John had to cut short his business trip and return to Texas. Within days of John's arrival home, his father passed away. He decided to stay with his mother and his sister, Paula, to make the necessary arrangements. At his father's funeral, John delivered the eulogy, and after the burial ceremony, the family returned home for a traditional Texas reception. Barbecue, casseroles, and pies streamed in. As the town's beloved postmaster, Joe Hanke had known all 893 residents of Cross Plains and they seemed to all be there to pay their respects.

In the middle of the reception, John's phone rang. He paused while studying the number. *Oh my god,* he thought. *Am I going to have to deal with this?* It was Silicon Studio, calling from Japan. John knew that he couldn't delay the deal any further, because Keyhole needed the cash. He excused himself from the reception, stepped outside, and negotiated the deal to fund a few more months of Keyhole's payroll, desperate to keep the company alive. John would later admit to me, "I had to wrap up the deal at my dad's funeral. That was a pretty low point for me. I was so spent. I had nothing left."

After John returned to Mountain View, he directed his attention to Digital Globe (DG), as they had recently launched their new QuickBird satellite. With Nvidia's appeal to consumers globally and a new distribution partner in Japan, Keyhole needed to expand its imagery coverage internationally. Despite their comparatively low resolution, satellites still presented numerous advantages. First, they were always up there. The satellite orbited for twenty-four hours a day, seven days a week, three hundred and sixty-five days a year. This translated to *significantly* more current data. Secondly, QuickBird knew no international boundaries. It was just as easy to collect images in South Africa as it was in southern Arizona.

After over a year of discussions, Daniel and John were able to conclude a deal with Digital Globe that would give it access to large quantities of imagery covering the major cities of the world. Digital Globe saw Keyhole as the perfect partner for reaching new categories of customers—commercial ones and even everyday consumers wanting to simply explore the globe. This satellite imagery had historically been reserved for the U.S. military, and Keyhole represented a potential path to another market.

John asked Phil and the engineering team to re-architect the Keyhole back-end server and front-end client software to allow users to switch between different imagery databases. Soon we had two Earths: one featuring J.R.'s data and one featuring Digital Globe's. Users could switch back and forth between the two.

In late summer of 2002, the Digital Globe data started really rolling in. Often, sitting in our conference room in one meeting or another, I spotted the UPS or FedEx truck pulling up and heard Dede signing for the package, and Chikai and Daniel emerged from their cubes to see what we had been sent.

Chikai and Wayne Thai, a young image-processing assistant, would begin the two-week process of importing the data for a single city (often between forty-eight and a hundred and sixty DVDs), color-balancing, optimizing for streaming, and then pushing live. It was an arduous process, using Keyhole's Earthfusion software tool; it involved sixteen steps for a single city. Wayne—who was Vietnamese and raised in L.A., with a passion for Rollerblading and the Lakers—was often the last to leave the office, kicking off a "cron job" that set a workstation in motion overnight. The next morning, Wayne examined the results. The process frequently failed, and he started it over again, earning the nickname Second-Chance Wayne. Other nights, he slept on the couches in the office, monitoring a job. The team began adding data as fast as they could open packages.

One Friday afternoon John walked into Wayne's cube and asked about an update that he had expected to be completed. It wasn't—and the imagery was needed in order to make sales at a trade show the following Monday. John implied to Wayne that if he didn't get it done, he wasn't sure there would be a job for him on Monday. Wayne had visiting family in town and plans for the weekend. After John walked away, Wayne punched the wall of his cube. Mark Aubin stood up from his cube and said, "Wayne, you don't understand, what John means is if we don't get this done, *none of us* are going to have jobs on Monday."

Soon thereafter I remember demoing EarthViewer at another trade show. My potential customer was a young real estate entrepreneur who was considering purchasing remote beachfront property for a new resort development in Nicaragua. As he described the region, we jointly flew from space down to Nicaragua, then to the Pacific coastline, then a little farther south as he directed. As we got closer, I could see that there was actually a high-resolution image tile of the remote jungle of Nicaragua along the coastline. It was a single wayward image tile that Wayne had managed to process from Digital Globe.

My prospect had gone silent. I panned onto a pristine isolated cove of white sand and aqua-blue waters. His eyes widened, and he demanded that I zoom back out. Nervously he surveyed the surrounding area to see if anyone else had seen the undisturbed coastline. *His* beach.

"You're saying anyone can see this?" he whispered.

"Anyone with our software can," I replied. "Yes."

He purchased the software, and this kind of subscription sale (at $599 each year) began to add up for Keyhole. Afterward, he said, "Hey, let me ask you something. What if I wanted to take this image out? Could I pay you to do that?" I didn't know how to respond. I never contemplated someone wanting their property taken out of the database. This was the first time Keyhole was asked this question, but it certainly would not be the last.

As our database of international imagery from Digital Globe grew exponentially, we tried to make inroads into new markets: television broadcasters, nongovernmental organizations (NGOs), and military and intelligence prospects. While commercial real estate and the new consumer market were gaining traction, cash flow was running tight. It was common those days for John to emerge from long sessions with our accountant in a serious mood after studying spreadsheets and determining which vendors would and which would not be paid that week. Keyhole was standing over a kind of craps table, praying that at least one of these larger deals would hit.

Dave Lorenzini, our sales guy, laid the foundation for several long-term prospects. I can remember being at a government trade show with Daniel Lederman and being overrun with interested urban planners and city management administrators.

Dave was scheduled to work the booth at noon so that Daniel and I might rotate out for lunch. He was nowhere to be found, and our calls went straight to voice mail. Three hours later, he showed up with a story about having sneaked into the keynote presentation, where Secretary-General of the United Nations Kofi Annan was speaking. In classic Lorenzini style, he managed to show Annan a demonstration of Keyhole in the hallway after his speech. "Yeah, I grabbed Kofi in the hallway after his preso and showed him our stuff. He just about wet his pants!" Daniel and I could only shake our heads and laugh.

Start-up companies need guys like Dave Lorenzini. The less you manage them, the better. You just never know what they might stir up. For Keyhole, Dave stirred up quite a bit: He managed to find footholds in the NGO world and with a random collection of foreign governments from Yemen to the United Arab Emirates. Dave also started a conversation with CNN to use Keyhole's EarthViewer software on the air to cover various news events. With several opportunities, Dave started the conversation, but for some reason, he couldn't seal the deal.

In late August 2002, my first daughter, Isabel, was born three weeks early at Brigham and Women's Hospital in Boston. A few days later, I received a congratulatory call from John and stepped out of the hospital room in the maternity ward. "We're going to close the CNN deal soon," John said, attempting to put my mind at ease about Keyhole's dire cash-flow situation. "That's great," I said, trying to share in his positive excitement. "We've come off the $350,000 price tag to $250,000 to get the deal done. Lorenzini says that they're going to sign on," John continued. As I stepped back into the hospital room with Shelley and Isabel, it was easy to recognize that the pressure was mounting.

Dave and Daniel had been working on the CNN deal for six months. Many smaller broadcasters and local stations across the country had signed on as Keyhole customers in order to use the software on the air to show traffic accidents, crime scenes, and other local breaking news stories. EarthViewer could almost replace the local news helicopter, which cost the average local news station something like $2,500 each time it was deployed.

By October, even with modestly growing revenues, Keyhole had burned through the Nvidia and Silicon Studio infusions. The company was surviving payroll to payroll, and investment prospects remained grim. As a Hail Mary, John and Brian put together a "friends-and-family" investment round. Each canvassed their contacts and reached into their own pockets to throw a lifeline to Keyhole. I still had over $10,000 in unpaid four-month-old expense reports, so I certainly could not invest. Altogether they raised approximately $500,000.

As a part of this round, Noah Doyle, a business school classmate of John's, came on board as an investor and an employee working on business development and partnerships. I liked working with Noah. Bright, quiet, and almost professorial, he often paused and ruminated on the options when asked a question, considering all possible sides of an issue before methodically answering. A serial entrepreneur, Noah also told John that he was willing to forgo any salary in exchange for equity in Keyhole. It wasn't his first smart investment bet. With another Berkeley classmate, he had cofounded a loyalty rewards program software company and sold it to United Airlines in July of 2001.

Later, Noah would prove critical to the company's business-to-business strategy and sales activities.

During the fall of 2002, a deal even larger than CNN was on the horizon. This sales opportunity centered on the use of our Keyhole software, but not our data. Instead, our software would utilize the customer's data—enormous databases of real-time, high-resolution satellite and aerial imagery that dwarfed the imagery databases we had patched together.

It happened like this: In the aftermath of September 11, Keyhole began to receive inquiries from various agencies in Washington, D.C. Tensions had been escalating in the Middle East, and it was clear that the military and other agencies were on the hunt for new technologies that might help with this effort. Keyhole began discussions with the National Geospatial-Intelligence Agency (NGA), the organization that generates global maps for the military. They had unlimited maps and data, but their mechanisms for distributing this data to the individuals who needed it were clunky and slow compared to EarthViewer.

Keyhole was contacted by an organization called In-Q-Tel. While all the other venture capital firms in Silicon Valley had essentially closed their checkbooks in 2001 and 2002, In-Q-Tel had actually opened up offices and was seeking out interesting companies to invest in. In-Q-Tel is not an ordinary VC; it is backed by funds from the CIA and the NGA, and had been created to help find new technology that might be helpful to the government. As we later learned, Keyhole was 80 percent of the way there toward being useful to their important work. In-Q-Tel's role was to pay start-ups, like Keyhole, to finish the final 20 percent of product features that could greatly benefit their work.

Unfortunately, this investment process was a predictably long one, and by the end of 2002, Keyhole's cash flow was tight yet again. Weekly staff meetings were sparsely attended and the engineers started showing up less frequently for work. Doug Snow quit abruptly, leaving Keyhole without a solid sales pipeline. CTO Phil Keslin left for a more stable job at Nvidia. Even I was looking for job possibilities and went on an interview; my conversations with a Cambridge-based company got as far as salary negotiations.

The consumer version had improved our revenue by about $50,000 per month, but that revenue was closely tied to the amount of exposure Nvidia was giving us on their website—and that audience had begun to dry up after the first few months of the deal. In our weekly calls, I often leaned on my Nvidia contact, Keith Galocy, for new ideas to help us promote our Keyhole EarthViewer NV consumer product with monthly specials and promotions: their website, trade shows, newsletters, and the like. Still, sales of the NV version stagnated.

I asked John if we could approach Nvidia about breaking out of the exclusivity of our contract. With the exclusivity clause, the math was against Keyhole succeeding and spreading to a larger community of enthusiasts. "Are you kidding?" he replied. "The exclusivity [to run on Nvidia hardware only] goes for another six months. They paid us $500,000 for the exclusivity. Don't even bother." I responded, "Will you at least let me try?"

A few weeks later, Keith came over to our offices for a meeting on a warm Friday afternoon. I walked through the full financial picture, not sugarcoating our prospects and the real possibility of Keyhole's shutting down. Keith was sympathetic, but allowing us out of our exclusivity deal wasn't his call to make. After about a month of back and forth, with a concession to add access to a planet Mars database in EarthViewer NV and priced ten dollars cheaper, Nvidia agreed to let us launch a consumer version *not* specific to Nvidia hardware. Keyhole was allowed to market a product called EarthViewer LT, priced at $79.95 a year, which ran on computers without requiring a Nvidia graphics card.

For Keyhole, withdrawing this exclusivity would mean that we would have a much better chance of converting someone to a paying customer. If a hundred people landed on our website, instead of having a chance to convert only thirty-five of them to customers (only those with an Nvidia graphics card), we would have the chance to convert about ninety-five of them to customers. (We still had no version for Macintosh computers.)

It was a big win for us, and John was appreciative of my efforts on this negotiation. But truthfully, the sales of the consumer version had been driven by the traffic from Nvidia's website. We did not have an equivalent channel of promoting this new consumer version. It wasn't as if we were, now that we had this widely available software, going to be able to start running television ads showing it to people.

Or were we?

In January of 2003, the CNN deal to license Keyhole continued to sputter along with little indication that it would ever materialize. It was a major disappointment. David Lorenzini had been working on the CNN deal for over six months, and Brian

McClendon worked on his childhood friend who happened to run CNN's graphics. In sales pipeline reviews, the opportunity had shrunk from $400,000 with a 95 percent likelihood of closing to $75,000 with a 50 percent likelihood of closing; it often generated frustrated glances whenever it was brought up during staff meetings.

Then David Lorenzini left Keyhole abruptly. He returned to his safer, more predictable paycheck at his former job. The CNN deal was almost forgotten. The last offer was that CNN would license Keyhole software for $75,000 per year and be allowed to use it on the air. With Lorenzini gone, no one was around to chase the deal to a close.

Between the commercial real estate market, the consumer version, and a hodgepodge of other sales, Keyhole had managed to generate $2 million in revenue in 2002. While that was up significantly from $500,000 in 2001, we still had lost money for the year. The initial capital investment from Sony was long gone, and the only reason we were still in business was because John had managed to patch together investments from Nvidia, Silicon Studio, and a friends-and-family round of capital.

By December of 2002, Keyhole was effectively broke again. The board was meeting at the beginning of January, and John was faced with the dim prospects of laying people off or shutting down the company. One night over a beer at the Sports Page, a popular Silicon Valley dive four blocks away from our offices, John and I discussed what might be done. During recent weeks, he had kept his chin up for the company, but when the two of us were alone, John appeared more deflated about Keyhole's future. That said, he still didn't let on just how bad things were.

What I didn't know at the time was that he had been personally lending the company money on occasion to meet payroll. The loans were repaid when revenue came in, but it was a contentious point with his wife, Holly. They had two young children, and although John's earnings from his previous companies provided some cushion, he was feeling the financial pressure, too.

John glanced up at the 49ers game on the old television mounted above the musty bar. He continued to pump me for ideas for short-term promotions in order to convert trial accounts to permanent subscriptions. He wasn't going to give up.

"How about the entire company takes an across-the-board fifty percent pay cut?" I suggested.

John stared silently across the bar, seemingly crunching the numbers in his head.

"Well, that would buy us a couple more months," he finally said. "Maybe enough time for the In-Q-Tel or CNN deals to come through. And maybe we could make up for it by giving the employees a little more equity."

That night John drove home. In the middle of the mile-and-a-half Dumbarton Bridge (which connects Silicon Valley to the East Bay), his car started sputtering, and he pulled over to the bridge's narrow shoulder. John glanced down at the dashboard: He was out of gas. John had to call Noah to pick him up.

Later, in January of 2003, in preparation for the board meeting, John asked me to help him with a bit of street theater for the benefit of the attendees, who were considering the option of shutting down the company. Our online sales had been steadily growing and it had been a common practice for me to send a monthly price promotion to all of the people who had signed up for a free trial of our software via our website. This was always good for a quick hit of revenue, but something that could be done only once a month. John asked me to tee up a special price promotion and hit send on that month's batch of free trials timed to hit people's inboxes right at the start of the board meeting in our conference room.

John's computer was set up to receive order notifications from our payment processor anytime someone purchased a license. So while John presented the dire cash-flow situation to the board, his presentation was frequently interrupted by notifications of orders for a Keyhole subscription. Now, I won't say that this little stunt saved the company, but John did smile wryly after the meeting and said with a wink, "It certainly didn't hurt the mood of the room."

The board decided to not shut down the company. Instead, each employee was given a choice. A sliding scale was created that would award equity to people who agreed to a temporary reduction in pay. It ranged from 10 percent to 100 percent. Most of the company agreed to take a significant cut. Some employees, like Chikai, were willing to forgo their salaries entirely in exchange for more equity. Expense reports remained unpaid. Countless calls from J.R. and other various vendors were avoided. We had a plan for one more quarter and we hoped that one of those large deals with CNN or In-Q-Tel would come through. We had three more months before John would have to shut down the whole idea.

Chapter 5

SITUATION ROOM

On the morning of March 27, 2003, David Kornmann was the first one in to work. He opened up the office and made a pot of coffee. As he walked by the fax machine, David noticed a fax had come in. It was a signed contract from CNN, curiously signed and sent by their lawyers back in Atlanta at three o'clock in the morning.

That afternoon John called to tell me about the $75,000 contract. His excitement was tempered, given the limited revenue of the deal. That said, John was using every bit of good news to try and keep the drained Keyhole team energized. He also mentioned something that he had inserted into one of the last proposals to CNN.

"I got them to agree to put our URL up on the screen anytime they use our software on air," he added. I acted excited for John's benefit. He had been persuaded to accept the lower price in exchange for this on-screen attribution by a former MBA classmate, Omar Téllez. Both Omar and John had young families and socialized together often. After a dinner at John's house, when he was describing Keyhole's struggle and the challenge of closing CNN, Omar suggested that they simply get on-air attribution.

Later that evening, CNN's twenty-four-hour coverage of the

U.S. invasion of Iraq opened a new segment at eight o'clock, with Wolf Blitzer bursting onto millions of television screens.

"The latest round of detonations came about an hour ago, more unfolding even as we speak. But there were perhaps even the largest explosions earlier tonight," Blitzer stated. "Witnesses reported the bombs struck several sites previously targeted in Baghdad, including a presidential palace and military positions on the city limits."

The segment returned to Atlanta with anchor Aaron Brown, who was stationed in CNN's Situation Room. Brown pivoted to reporter Miles O'Brien, who was, unknowingly, about to change the financial trajectory of our small mapping company back in California.

Brown segued to O'Brien: "Miles O'Brien is at the map table, I believe, and has more on why they would go after these sites. Miles—"

I should point out that the plan for using Keyhole on CNN had always been that we would work with their graphics team to create animations: prepackaged videos that could be played occasionally to supplement the coverage. You've seen these map animations; they typically make up three to eight seconds of a thirty-minute news program and provide the viewer with a bit of geographical context about an event.

Miles O'Brien had other ideas. Instead of creating short prepackaged videos, he planned to fly free-form around Baghdad for lengthy feature segments. He was essentially going to turn CNN, with its millions of viewers, into a Keyhole EarthViewer show-and-tell.

Years later, O'Brien told me: "I was just cocky enough to think that I might pull that off. No one else at CNN would have tried to do it, but I knew it was going to be a hit because during breaks in the broadcast, there would literally be a line of people inside the newsroom at my computer wanting me to fly to their house in EarthViewer."

When Brown transitioned to him, O'Brien commenced his Keyhole EarthViewer high-wire act with a degree of trepidation: "Aaron, we're actually taking a look at some satellite imagery, which is rather dramatic. First of all, I want to tell you a little bit about how we were able to capture this imagery."

CNN floated a perfectly detailed rendition of Planet Earth with *EarthViewer.com* displayed prominently in the upper-right-hand corner.

O'Brien zoomed from space down to Baghdad, demoing for millions of CNN viewers worldwide our Keyhole EarthViewer software. With the updated Digital Globe satellite imagery that showed recent bomb damage, O'Brien implied that the image provided an almost real-time view of the war action.

"You notice the time here, that's Universal Time, Greenwich Mean Time, if you will, 7:35 A.M. today. That satellite, at about a hundred miles above Baghdad, captured some images of Baghdad which are very telling. As we zoom in on Baghdad, using Keyhole's EarthViewer software, we'll show you exactly what we're talking about here. We can sort of do our own armchair bomb-damage assessment."

Meanwhile back at our office, Chikai jumped up from his desk and hustled to Ed Ruben's cube. As head of Keyhole's server infrastructure, Chikai had weathered spikes in traffic before as we gained ground with commercial real estate brokers.

Initially, the Keyhole team had purchased an expensive RAID storage array, but as the company's data and needs grew and its resources dwindled, the engineers started buying racks, boards, and disk drives configured with Linux from Fry's Electronics. The company's growth had been punctuated by regular trips to Fry's to stock up on discounted drives and server boxes to add capacity as more users and more data were added to the system. Chikai's patchwork of low-end Linux servers had always managed to stay one step ahead of demand.

This spike was different. More of an explosion than a spike, it interrupted the otherwise-quiet Thursday afternoon. Measured in fifteen-second increments, the server traffic load shot up exponentially. Then there was silence. Then all Keyhole EarthServers were down.

Or all but one server. CNN's version of Keyhole EarthViewer was pointed at a separate server, dedicated to just CNN. O'Brien continued: "We're going to zoom in on Baghdad and show you this new animation, this new imagery that we just got."

Back in Mountain View, Ed and Chikai didn't know what was going on. News of the CNN deal signing had circulated throughout the skeleton crew at Keyhole that day, but the work to create the CNN animations hadn't even started yet.

"I'll try rebooting the auth server again," Ed said to Chikai. Ed had kicked this can down the road before. While he had raised the point that the authentication server (auth server) code base was in need of a complete rewrite, it had never risen to the level of importance to get done, due to limited resources at Keyhole.

Besides, simply rebooting the auth server—the system that created and authenticated new user accounts when someone wanted to subscribe to Keyhole EarthViewer—had always worked before. It was the company's cash register and turnstile, an old and persnickety machine that controlled all access to our service. This single server determined who could and who could not come in the door to use Keyhole EarthViewer.

"Guys, we're down," said Norm McClendon, the burly yet soft-spoken head of technical support for Keyhole and also Brian's brother. He had joined Ed and Chikai in the cramped, overheated server room, which was more closet than room. Norm's support lines were ringing endlessly as Ed tried for the third time to reboot the auth server. Each time it restarted, however, the crush of traffic overwhelmed the cobbled-together machine.

"Guys, what the heck's going on?" John yelled out from his cube. "We're down." There was no need to remind anyone of the revenue implications of every minute the company's subscription-based service was offline. Outages often translated to subscription extensions, and sometimes refunds.

Keep in mind, we didn't have a television hooked up to cable and CNN. The only television in the office, which lived in our small conference room, was used mainly for late-night Gran Turismo racing sessions on the Sony PlayStation while an engineer waited for a new imagery database to be processed. No one knew what had caused the traffic spike. Chikai looked down at his phone as John came into the server room and read a text from a friend: DUDE! I JUST SAW YOUR COMPANY ON CNN!

The next day John excitedly said to me on the phone, "I think you better get out here. Why don't you see if you can catch a flight?" His tone was hurried and breathless. This was my first sign that our financial outlook might have just changed, because recently I had been told to discontinue my monthly trips to Mountain View.

Back in Boston, I was officing out of a coworking space owned by Bill Warner, founder of the leading software video-editing company Avid. Warner had turned his attention to fostering early tech ventures in Massachusetts, and I was lucky enough to be one of the twenty-six people renting out a cube in his building.

Bill was the first one to congratulate me when I walked in the day after CNN started using Keyhole on the air. He dropped that morning's edition of *USA Today* on my desk and said, "Well, I guess you'll be moving to California." The headline read TINY TECH COMPANY AWES VIEWERS. Kevin Maney reported on CNN's usage of Keyhole EarthViewer and the subsequent overwhelming demand: "On Thursday, CNN used the maps to simulate flying over Baghdad and dropping down to street level at bombing targets. Each time, CNN flashed the website, EarthViewer.com . . . users overwhelmed the site, bringing it down for most of Thursday and besetting Keyhole's twenty employees. Late Thursday, CEO John Hanke sounded frazzled, though he noted, 'There are worse problems to have.'"

Articles appeared in almost every major newspaper and magazine. *Newsweek, Time, The New York Times.* But this publicity seemed almost trivial, a mere blip of attention compared to the exposure that was playing out 24-7 on CNN. Truthfully, no one in the company at the time recognized the value of that little concession to show our website address on the air. Not me. Not John. And certainly not Dave Lorenzini, who had worked to make the CNN deal happen only to leave a few short weeks before the U.S.-led invasion of Iraq. Nor did we think that Miles O'Brien would use it live on-air for extended reports about the bombing campaign. Nor that the war would last for years, costing the lives of over five hundred thousand Iraqis and over five thousand U.S. soldiers.

O'Brien continued to use EarthViewer for extended segments, even though the software occasionally crashed while on the air. "And what I want to point out to you is, as we zoom in a little bit closer on this, it is quite clear that in this area, which is obviously wooded, and General Don Shepperd, maybe you could help me out and just point out in the middle there, there's at least one *tank*. We spotted several other tanks, we think, near the trees, and the associated treads with it."

I caught the earliest flight out from Boston. That morning, when I walked through Keyhole's front door, Dede stood behind the desk. She looked up from a stack of papers in her hands. Her blond hair was tousled, her eyes weary. She peered over her reading glasses. It was ten-thirty in the morning, but Dede looked as if she had been working for the past ten hours. The phone on her desk rang, but she didn't answer it. In fact, it sounded as if every phone in the office was ringing.

"Well, it's about damned time!" she barked at me.

The phones continued to ring. Truthfully, unless you knew someone's cell phone number, you weren't going to be getting ahold of anyone at Keyhole.

Special shipping crates were splayed out on the floor near the pool table. Lenette held a checklist of items being prepared for immediate shipment, and she was printing, binding, and inserting CD-ROMs into multiple copies of manuals. Much of the documentation was being typed up, printed, and immediately inserted into the two-hundred-page instruction manuals.

"What's all this?" I asked Lenette.

"You'll have to ask John," she said as John walked up to the chaos.

"You're here!" John said, happy to see me if not a little surprised. Amid the CNN frenzy, he had forgotten his urgent request for me to return to Mountain View.

"What's all this?" I asked again, still confused, looking down at the boxes being packed with DVDs, hard drives, user manuals, and machines.

"I'll tell you about it later," he said, spry and energetic. "Let's grab lunch today."

That afternoon John told me that Rob Painter, a partner with In-Q-Tel, had managed to get through to him on his cell phone. After months of bureaucratic back-and-forth, the In-Q-Tel deal was finally signed. At $1.5 million, it was by far our largest contract to date and enough to fund Keyhole's operations for six more months. Those who had dropped their salaries received makeup payments and everyone returned to their original salary levels. With a maturing product, growing revenues, and working capital, Keyhole was finally on solid ground.

An aggressive timetable of project deliverables was a part of the contract. This In-Q-Tel schedule was accelerated since the entire State Department, every U.S. senator, every congressman, every ambassador, every foreign government, every military intelligence commander, the secretary of defense, the Joint Chiefs of Staff, the entire military-industrial complex was witnessing the twenty-four-hour nonstop infomercial for Keyhole Earth-Viewer software as demoed by Wolf Blitzer and Miles O'Brien. And many of them were asking the same critical question: *"Why the hell don't I have that on my desk?"*

"Who is asking?" John asked Painter. The whole team was, of course, curious about the final destination of our Keyhole EarthSystem. The team speculated that it was Secretary of Defense Donald Rumsfeld, Vice President Dick Cheney, or Secretary of State Colin Powell.

"You know I can't tell you that," Painter replied, knowing of John's past career in the foreign service. "Honestly, it would be easier for me to tell you who is *not* asking." Painter, a former Special Ops soldier who still looked like he could strip off his sports jacket and jump on a C-130 to Afghanistan at a moment's notice, had a simple message for John: *stop.*

Stop everything we were doing. Stop all development work. Stop all consumer work. Stop the technical support and imagery updating for CNN. Instead, focus on getting a private version of the Keyhole EarthSystem, including all of our data, software, and hardware, for use inside the U.S. government. The "representatives" would be arriving at the Keyhole offices on Wednesday at five o'clock. Everything needed to be crated and ready to ship by that time. Painter realized this would mean that the price of our contract would go up. He didn't care. Price was not an issue.

Keyhole was shipping an entire version of our systems to them: servers, boxes, CD-ROMs, and all of the necessary documentation. Some of the last software scripts that controlled the imagery import process were being written, burned to CD-ROMs, and going directly into the crates. There was no time for any quality assurance testing. The CIA and NGA were counting on this Keyhole system—they were going to be importing massive amounts of their own satellite imagery, overhead surveillance photos of the war theater of much higher resolution than anything we had access to.

The entire team, including John, was busily packing and taping and boxing and shipping all the while with the TV, now with cable service, broadcasting CNN, often with Keyhole being used on the air. "We're on again," someone yelled, and we'd all go scrambling to the conference room to watch another CNN segment with Keyhole. It was happening so often, it was becoming less novel and exhilarating. The phones continued to ring, going unanswered as hundreds of people were calling and simply pressing any extension in the hopes of getting through. Our consumer and commercial servers were often knocked offline as Ed and the team tried to rewrite our authentication software. Brian made a late-night run to Fry's and bought twenty-three hard drives and stuffed the equipment into the small trunk of his yellow Mercedes sports car (the manager had to approve the purchase because customers were limited to buying ten at a time).

As the crates were packed up, John joked that we should probably ship Chikai in one of the boxes, too. That call would come later. They wanted Chikai, too. First to Langley and then to the National Geospatial-Intelligence Agency in St. Louis, considered to be the Fort Knox of surveillance satellite imagery. Without the proper clearances, Chikai wasn't allowed to set his fingers on a keyboard at Langley, but he could stand behind someone and instruct them on the necessary Keyhole EarthServer commands as they loaded up their version of the software using their own satellite imagery. Lenette went with Chikai to help with the training sessions for In-Q-Tel and their clients in Washington, D.C.

For the government, military, and intelligence communities, Keyhole proved to be an incredibly powerful technology. It allowed them to take tens of thousands of photos of the real-time war theater, stitch them together in one giant mosaic, and then stream that database out over a network to thousands of military and intelligence officers.

We started hearing about individual soldiers and their anxious families signing up for Keyhole on their own, not waiting for a controlled briefing or government procurement. They were signing up for our free trials and standard subscriptions (though we were prohibited by law from allowing downloads to anyone in Afghanistan or Iraq, which became a problem with soldiers trying to download from the war theater). We heard from multiple soldiers who were signing up to learn the terrain where they were being deployed. In one anonymous voice mail left for the company after hours, an "operator" had called to say thank you. He and his team had come under fire and needed to figure out a way to safety. Deep in enemy territory, they had resorted to EarthViewer, connected via a satellite data connection. With new insight into their location and the surrounding environment, they had been able to find a way out.

Keyhole ended up taking satellite imagery of the war theater from Digital Globe and putting it on the desktops of thousands of military, administration, and political leaders. It allowed them to see the imagery and terrain for themselves, instead of waiting for imagery maps created by the GIS experts. More important, the constant exposure on CNN had shown millions of people outside of the military what this new war looked like through the lens of Keyhole. Interest poured in from all kinds of businesses and consumers. At only $80, individuals could plan vacations and conduct real estate searches with the power that a few years earlier had been available only to organizations willing to invest millions in high-end supercomputers and satellite data.

We spent the spring of 2003 trying to respond to the flood of demand. As perverse as it sounds, the invasion of Iraq proved to be a turning point for Keyhole—a huge cash injection when we needed it most. Though I was adamantly opposed to the invasion, it actually helped the company I worked for financially. As I walked John through a recent sales report during our Monday-morning staff meeting, he commented, "Well, I guess we'll be able to make payroll."

Throughout the early weeks of the war, we continued to provide CNN with updated imagery of Baghdad. Using a new feature called image overlay, O'Brien could on the air use a slider to show what one of Saddam Hussein's many palaces looked like one day and what it looked like the next, sliding the transparency of the new image back and forth to show what had changed.

The image overlay used by CNN to cover Baghdad was a basic—if visually stunning—trick that military intelligence image analysts using Esri software had access to, but not something that the world had seen before. It made it easy to visualize what a site looked like before and after something had happened, be it bomb or hurricane or other disaster.

Through the summer of 2003, CNN continued to use Keyhole to report on the ongoing search for the alleged stockpile of "weapons of mass destruction" and other newsworthy events.

Unlike the early days of the war, we were able to keep the servers up, though Chikai resorted to carrying a pager in order to closely monitor the servers at all times.

Our conference room was now our own situation room. We set up two screens: on the left, a TV with CNN playing nonstop, and on the right, a computer monitor with an internal dashboard showing multiple server metrics, such as unique visitors currently on our site, trial downloads, user sessions, and sales orders. On the television, O'Brien started again: "And we are getting reports, and you can see here using Keyhole EarthViewer." Then the monitor with all of the metrics on the downloads, bandwidth, and purchases refreshed every fifteen seconds, showing the corresponding jumps in sales.

We found it intoxicating to watch—a small company on such a huge stage—and often loitered in the conference room for long periods of time. One day John cut us off. "Guys, let's get back to work," he said, clapping his hands together. He reminded us that while this was all great, we still had lots of work to do.

Keyhole had been in conversations with several other media outlets, and once the war had started, they called John and Daniel to sign on. *ABC News* was the next to go online, and Peter Jennings began using Keyhole EarthViewer. Al Jazeera signed on. The BBC signed on. CBS expanded its usage, including *60 Minutes*. The broadcast business opened several other avenues—from an ESPN bass fishing show to the Tour de France to local news affiliates all over the world. The broadcast business ended up being an unplanned boon to the company.

In the spring of 2003, broadcast television and consumers generated a critical cash influx for Keyhole, but commercial real estate customers continued to be our foundation. Since Keyhole was an annual-subscription-based business, customers from 2002 signed back up for another year. These renewals made for comparatively easy money. I can remember a check for $14,000 from a commercial real estate company, with a site license for over twenty users, showing up in the mail one day. Due to a data input error in Salesforce, we had not noticed that their deal was up for renewal. But they had dutifully sent their check in anyway, without Keyhole even making a phone call.

Back in Boston, I told Shelley, "You know, when checks for $14,000 start showing up in your mailbox unexpectedly, that's a pretty good sign that the company is onto something." She agreed. After much rumination, and a conversation between Shelley and Holly Hanke (who was optimistic, but still cautious), we decided to finally make the move that John had wanted us to make four years earlier. In April of 2003, we packed our bags for California. We were all in.

Chapter 6

BEHIND THE CURTAIN

"I want to get everyone's attention," John said, rolling up the sleeves of his untucked blue button-down and glancing at his folded-over handwritten notes. "I'll make this short, but I do have a few introductions to make." For John, he looked relaxed. It was a breezy Friday afternoon in June of 2003, and the Keyhole team of twenty had filtered onto the small grassy lawn, reclining on cheap plastic chairs under the shade of a giant eucalyptus. Mark Aubin manned the rusty propane grill, cooking up Polish sausages and sauerkraut. Wayne Thai pulled out a cooler of beer, and Dede Kettman spread out paper plates, assorted condiments, and Costco chips and dips on a foldout table, which was a freebie found on Craigslist. (It's likely all of Keyhole's furnishings and decorations were Craigslist finds.)

I cracked open a Sierra Nevada Pale Ale, spread out on the grass next to Ed Ruben, and clinked his beer. "Nice job this week, Ed," I said for no apparent reason. There were many reasons to raise a glass those days. While we weren't completely out of the woods, we could sort of see the path. Life at Keyhole was totally different from four months ago.

The In-Q-Tel earn-out milestones were being delivered: We were being paid to make our software better. CNN and other broadcasters continued to drive consumer exposure for Earth-Viewer. Our core commercial real estate customers were signing on at a strong clip, with many of them committing for their third year of subscription.

Keyhole's EarthViewer software had only improved with time: more stable, more useful features, and with access to more imagery data. Add to all that the fact that just as Moore's law had predicted, the percentage of computers now capable of running EarthViewer had doubled every year in the four years we'd been operating.

John tapped his beer with his car key, trying to get the team's attention. "Let's get started. We've got some new folks that we want to welcome to the team. Bam, why don't you come up here." It was an odd role reversal for John to bring up Brian A. McClendon (or Bam, as he was called), Keyhole's new VP of engineering. After all, it had been Brian who had found John to lead the Keyhole spinout. Brian had been on the Keyhole board of directors since its inception. John had had to talk Brian into taking on this role after Intrinsic closed its doors.

Three years earlier, it had been Intrinsic with the cash that helped the floundering Keyhole. Now it was Keyhole with the cash. Intrinsic Graphics, despite being stocked with an accomplished group of engineers, had never been able to find its market. Game developers wanted more control over their high-performance code, and Intrinsic hadn't been able to convince them to adopt its elegant but limited solution in large enough numbers. Despite raising over $15 million in VC funding, Intrinsic was dead. And with green fields of opportunity in front of us, Keyhole would prove to be a soft landing for many of the company's best and brightest. John was in a position to win the services of the most talented engineers from Intrinsic.

He hired Michael Jones, the other cofounder of Intrinsic, that summer, too. Michael was a true pioneer in the world of 3-D computer graphics. While at Silicon Graphics, he had helped to create an application programming interface (API) for controlling the rendering of 2-D and 3-D graphics called OpenGL. (This set of software functions is at the core of how graphics processors, like Nvidia, are controlled.) Technically gifted, the owner of two dozen patents, and a brilliant public speaker, Michael dropped out of college after one year, teaching himself advanced mathematics and computer science. When asked where he attended school, Michael offered only John Adams Elementary School in San Diego, California. He is, quite likely, the smartest person I have ever met.

John Johnson joined Keyhole and, with Mark Aubin and Chikai Ohazama, rebuilt the entire Keyhole Earthfusion data processing tool set. This became the central factory that all imagery and geodata traveled through; basically, the core system took millions of individual satellite and aerial images and stitched them together into one seamless color-balanced mosaic while optimizing the images for streaming. Originally created by Phil Keslin, this server-side innovation was the true patented secret sauce of Keyhole that allowed us to stream a single enormous imagery database to tens of thousands of subscribers.

And in what everyone on the Keyhole team considered to be a major coup, Brian and John had managed to snag John Rohlf, a software engineer formerly with SGI who is still considered among the best 3-D graphics super programmers in the world.

At Keyhole, Rohlf went to work on a new EarthViewer client application. He was both revered and a bit feared for his technical prowess.

I knew John Rohlf by reputation only. Unshaven and quiet, he walked slowly around the office in his worn-out Birkenstocks with a baseball cap pulled over his eyes, and seemed to be always lost in deep thought. John came and went at odd hours, with his laptop stowed under his arm, walking right past me and others, not saying a word. We pieced together details of his background like a jigsaw puzzle. He was from North Carolina, rumor had it, where he had studied computer science at UNC-Chapel Hill. From what I could tell, the only money he spent was on drag race cars and the fastest, latest Mercedes on the market, no matter the cost, which he drove in flip-flops, shorts, and threadbare T-shirts.

Weeks later, I sensed someone pull up next to me at the urinal at the Keyhole office. I glanced ever so slightly to my left. Yep. Rohlf. As we both washed our hands at the sink, John glanced up at me in the mirror.

"You don't understand how big this will be, do you," John Rohlf said to me. Was he actually speaking to me?

"What?" I asked with some trepidation.

"KML."

Of course I had heard about a requirement of the deal with In-Q-Tel from Lenette, the Keyholer who was managing the project. A new software standard—called Keyhole Markup Language, or KML—would enable users to share points, lines, and polylines with other Keyhole users. User A could draw on his or her EarthViewer and then send those annotations and drawings to another EarthViewer user B as a KML file, and user B would be able to see the exact same view of the Earth, with all the same annotations and the same perspective of the Earth. It was going to be powerful stuff.

"No, I guess I don't know, John," I responded, grabbing a paper towel.

There was silence. I looked up at John in the mirror. His ever-present baseball cap concealed his gaze. He finished washing his hands. Snatched a paper towel. Crumpled up the paper, looked at me in the mirror, almost disgusted by my lack of understanding of the potential implications of this new standard for sharing geographic information. He threw the crumpled-up towel in the trash with a bit more force than normal. And walked out the door.

And that was that. KML would arguably become the single most important improvement to our software, changing Keyhole EarthViewer from a single-user application to a collaborative one that could create and share geographic information.

With Keyhole, a real estate broker could measure the distances between competitor locations, the acreage of a plot of land, or even the square footage of a warehouse by simply drawing a line around it. And he or she could mark up that map with arrows, dots representing a location, or the outline of a property.

Now, with Rohlf's KML, those measurements and annotations were not just on one computer; those measurements and annotations could be shared with other Keyhole users. KML became the standard by which Keyhole users worldwide shared their data, their view of the world. The .doc of Word or the .xls of Excel. Eventually it was used by millions and was created by John Rohlf, a genius working, essentially, in total isolation.

McClendon, Jones, Johnson, and Rohlf were disappointed that Intrinsic Graphics had not survived. But for Keyhole, we were happy to double the size of our tech team. With the addition of these new resources, Keyhole had the ability to take on new types of work and industries, including visualizing and sharing data from GPS satellites using the new KML data format.

The Global Positioning System constellation of twenty-four satellites orbiting the Earth had been launched for military purposes in the 1970s, starting with the need to locate submarines and other military assets. In 1991, the Gulf War was the first military theater to rely extensively on GPS, though the receivers required still weighed over fifty pounds in the early 1990s.

In 1996, President Clinton directed the military to stop the practice of degrading the accuracy of GPS data for nonmilitary devices. The GPS satellite signals were opened up for consumer uses. These signals were made available by the United States worldwide, free of charge, and gave birth to a new type of consumer electronic device, the personal navigation GPS receivers. Though the consumer devices cost hundreds of dollars, the size of a receiver had shrunk to under three pounds by 2000.

Among the many uses of this new GPS technology was law enforcement. Police forces across the country began to take note of the decreasing size of GPS receivers and cheap new GPS tracking devices soon became commonplace. Have a suspect in a case that you want to know the whereabouts of? Simply sneak up on his driveway and slap a magnetic GPS tracking device under the bumper of his car. Or on his gold Ford F-150 pickup truck. Which is exactly what the district attorney for Santa Clara County did to Scott Peterson, the chief suspect in the killing of his pregnant wife, Laci.

Peterson wasn't being tried in Santa Clara County, but the district attorney worked on the case nonetheless. He had developed a reputation as a thought leader in the use of GPS and mapping technologies in order to solve crimes. He came to Keyhole in the summer of 2003 to meet with John Hanke and me. He had GPS data from Scott Peterson's truck, and the D.A.'s office was a subscriber to Keyhole EarthViewer. By that time, a dozen local police forces had become Keyhole customers because they were tired of waiting on Esri map screens to reload a mapview.

The Santa Clara district attorney's office had closely examined John Rohlf's new KML standard and early map annotation efforts. In the app, we called them EarthViewer placemarks. More specifically, these structured data files included a latitude and a longitude, an elevation setting, and a viewpoint angle. These placemark files told EarthViewer where to fly and what to show on the screen. EarthViewer was no longer just a map; it was a map that you could draw on and make your own.

This KML placemark data structure was not that different from the GPS data that the D.A. had captured from the GPS on Peterson's truck: though that GPS data also included a time stamp showing the time each latitude/longitude location had been visited.

John agreed to take some of the sample data and see what we could do. He delegated the task to a new Keyhole engineer named Francois Bailly. Importing the latitude and longitude and turning them into distinct placemarks was child's play to Francois. He completed this within a few hours. The D.A. would be able to manually click his way through the various points. But Francois was interested in the time stamp: The GPS data showed the time for each latitude and longitude combination.

With some trivial math (calculate the distance between two points, calculate the difference in time, divide the distance delta by the time delta), he could measure *speed*. And if *where* Scott Peterson had traveled didn't implicate him, the *speed* at which he traveled certainly would.

In the weeks following Laci Peterson's disappearance, Scott Peterson had emerged as the leading suspect. But months into the investigation, there was still no body, complicating the case against him. The Modesto police had slapped the GPS tracking device on his pickup truck about two weeks after the disappearance. In the weeks that followed, Peterson, the GPS data showed, had returned to the Berkeley marina, where he had supposedly gone fishing Christmas Eve, the day Laci had disappeared.

The GPS data—and Francois's alchemy—enabled the jury to see what Peterson had done when he returned to the marina. This is what they saw: an aerial view of the Berkeley marina in EarthViewer at a tilted perspective, and a red dot representing Peterson's truck in the foreground. The D.A. hit play. Peterson and the judge and the jury and the entire courtroom watched, in pained silence, the replay of Peterson driving up and down the shoreline of the marina at a very slow average speed for several protracted minutes.

Weeks later, the bodies of Laci and Conner Peterson, her unborn son, washed up on that shore. Jurors had to ask themselves: What could Peterson have possibly been looking for as he canvassed the shoreline at such a slow speed?

San Bernardino County in Southern California was another early adopter of our software, opting for a Keyhole deployment for its forest firefighting maps because of the speed and interactivity of the streamed, 3-D experience, and because it could be used by actual firefighters, not just GIS specialists. The fire department was willing to forgo the sophisticated analytical and map creation features of Esri for the speed and simplicity of Keyhole's EarthViewer.

What was particularly notable about the sale of Keyhole software to San Bernadino County was the fact that it was also the home of the Esri headquarters. It was a Keyhole shot across the bow of Jack Dangermond's multibillion-dollar digital-mapping empire, and one that John didn't mind pointing out to our team. I sensed John was growing in confidence as the leader of Keyhole.

John was eager to attract more of Esri's traditional customers by any means necessary. In the summer of 2003, we applied (and were rightly rejected) for a booth at the Esri user conference in San Diego, attended by twenty-five thousand Esri-trained mapmakers, consultants, integrators, and customers. We hoped to try to win away more core GIS customers, and we were also trying to learn more about a new Esri product that had been created in order to compete with Keyhole. Some of our customers and prospects had begun telling us earlier in the year about a new product that Esri sales reps called Esri ArcGlobe. They were pitching the new software as one that would offer the speed and interactivity of Keyhole, but with the sophisticated data analysis tools of Esri. Keep in mind, Esri still provided far more advanced tools than Keyhole: heat maps, threat domes, least distance calculations for fleet dispatch, and crime incident reporting and analysis. If Esri could combine its data analysis capabilities with a simple, fast visualization tool, EarthViewer was going to be in big trouble.

After hearing the news of our rejection, John decided to crash the Esri user conference. We were all surprised by this rebellious strategy. It was a brazen move by a CEO of a scrappy tech start-up. John asked me to go with him to San Diego on this guerrilla marketing mission. I had two boxes of snarky T-shirts—*Keyhole. GIS Gone Wild*—made and brought a couple of cases of EarthViewer demo CDs to hand out.

With our boxes of T-shirts and demo CDs, we wore trade show badges hanging from official lanyards; they just weren't from the Esri trade show. At the San Diego Convention Center, we managed to find a distracted security guard and bolted through. We were in. But now what? Neither John nor I had the personality of a true party crasher. For two hours, we walked around exhibitors' booths, sheepishly handing out our swag to wary Esri loyalists.

Despite our weak guerrilla marketing effort, we did manage to get our first look at the vaunted Esri ArcGlobe. It was slick, it was fast, it was beautiful. John was stone-cold silent. I said nothing as we both watched the demo in horror—for about thirty seconds. Then we got a peek behind the curtain of ArcGlobe: I noticed that demo was running on a machine with no network cables and not connected to a wireless network. Instead, it was running on a single machine, with all the data stored locally on its hard drive! The data was *not* on a server, meaning their 3-D model of the globe and all the data were limited to that single machine. When I asked the Esri sales rep about this, he dutifully answered, "The server infrastructure is still being optimized." (This is exactly the line I would have used if I were in his shoes.) The truth was this: Esri had simply figured out the easy client part, not the server part, which was Phil Keslin's true engineering leap forward, allowing Keyhole to stream a massive map of the Earth over the Internet. It was just another single-user tool from Esri, limited to the back-office GIS professionals, like every other Esri tool. ArcGlobe would never compete with Keyhole.

Later that afternoon, John left the conference to take a work call, and I walked around, carrying my box of dwindling T-shirts and CDs, and encountered another demo with equally dubious commercial prospects. The company's founder, a friendly Chinese national, manned a small table at the rear of the exhibition hall. Unlike most participants at the conference, he was excited to meet someone from Keyhole and to show me what he was working on. A Keyhole subscriber, he opened up EarthViewer and flew down to San Jose, California. He had a folder of KML placemarks represented as dots along a boulevard in downtown San Jose.

Then he showed me that if you clicked on a placemark dot on the map, an info bubble window popped open, and inside of that window was a picture. He told me that it was a photo of that exact location on the map taken from street level. I was amazed: I had never seen such a view.

Also inside the pop-up window, next to the photo, arrow icons pointed in different directions. He explained: "You can click on the arrows and it shows you a picture of the street in four different directions." The experience allowed you to fly down to street level in Keyhole EarthViewer, click on an icon on the map, and then turn in any direction to see the view of that location.

"How did you get those pictures?" I asked.

"I mounted four cameras with GPS on top of my car." He

pulled up a photo on his computer of his Chevy SUV with what looked like a disassembled mutant cyborg lawn mower on the roof.

"Very cool."

He explained further that he was going to continue capturing as much of San Jose as he could, sharing with me the number of miles of streets he had acquired, the months of driving he had done, the time, the gas, the equipment, the staffing needed, the cars needed, the software processing it, and the unknown market possibilities of such an offering. As the conversation continued, I worked through the math in my head. If he employed hundreds of drivers and cars around the globe, it would take many years to capture all of that data. And after you were done, you would have to do it all over again. Right away, it became apparent that this was one of the most economically ridiculous business ideas I had ever heard. It would cost literally tens of millions of dollars, no exaggeration. No, scratch that, *hundreds of millions of dollars* to undertake such a scheme.

I found myself trying to be positive about his prospects and to think of ways that he could potentially expand the concept. "Maybe a deal with UPS?" I suggested. Conceptually, this idea was similar to what Jen-Hsun had asked John about during the kickoff meeting with Nvidia—the notion of "going procedural" once you zoomed all the way in to street-level views. I wished him well on his project, accepted his business card, and walked away.

While our guerrilla marketing effort had been ill-conceived, the trip to the Esri user conference had been a productive one. Soon after, during the fall of 2003, Esri and Jack Dangermond approached Keyhole and John about partnering after realizing that Esri ArcGlobe was a technical dead end. His question for John: Could Keyhole act as a front-end visualization tool for Esri-created data?

The team began working on ways to work with Esri, including supporting the import of the Esri data file format, but John was also wary of getting too close to a big enterprise software company with large sales channels, system integrators, deployment schedules, and long government sales cycles. Given all of the other priorities and demands that Keyhole had in front of itself, John did not place a high priority on the custom-engineering support needed for Esri. His stance on the GIS market had not wavered from when I first proposed it to him in 2001: "I don't want to get pigeonholed."

In the meantime, during the coming months, John trained the focus of Keyhole's engineers and resources on the new version, Keyhole 2.0, slated to launch in early 2004. It was certain to be a major step forward for the company in terms of speed, measurement, annotation tools, and data. John continued to add talent to the Keyhole team to improve EarthViewer: Wes Thierry and Francois Bailly and his identical twin brother, Olivier. This team worked on forking the software, so they lived in two different codebases—one for professionals and one for consumers. I hired an intern named Jason Cain to import the digital versions of transit maps throughout the United States. By this time, these maps were being shared as KML files, which Jason could import into Keyhole. The city agencies were contacting us and saying, "Hey, we have this data." It was the first time that I fully understood what John Rohlf was talking about during our brief exchange in the bathroom.

In the spring of 2004, John appeared with Jack Dangermond at a military intelligence trade show in San Antonio. After their presentation, Dangermond sought John out. I strategically positioned myself to eavesdrop in on the conversation. It was big news to have Jack Dangermond, the creator of the entire digital mapping industry, in our booth. I was starstruck. Though he looked like a disheveled bureaucrat in his worn suit and tattered leather satchel, this was the man who created digital mapping.

"John, is there anything about our company that *concerns* you about working with us?" Dangermond asked, leaning over him like Lyndon B. Johnson. Clearly he was frustrated with the pace at which Keyhole was working on Esri's integration. It was likely that many of his top customers were asking about the delay, including the military-intelligence community. At the same time, Miles O'Brien was still using EarthViewer in his CNN broadcasts about the ongoing war in the Middle East.

"No, Jack," John said, laughing nervously. "We're working hard on it. I promise to get back to you with an update during the next two to three weeks."

What Jack didn't know was that he was too late: There was a wild card now in the game, and John was playing with a different set of cards entirely. A set that did not include Esri.

In those days, I occasionally rode my bike to work. As is generally true in Northern California, most days were beautiful, the temperature often a perfect seventy-two degrees. Starting in Palo Alto, I rode six miles on a well-marked bike path, crossing over the 101 and passing Intuit, the Sports Page bar, the Computer History Museum, Microsoft, and dozens of small tech start-ups. One such start-up was housed in a low-slung two-story building, which was recessed from the street amid tall redwoods. There was a small, nondescript sign in front with its name—Google. In 1999 (the same year Keyhole was founded), Google had curiously entered a market that most considered oversaturated. Every time I cut through the company's parking lot, there were more and more cars jockeying for spots.

I remember wondering if maybe another company had moved in to the building and just hadn't put its sign up. I finally decided to skip the shortcut, steered clear of the overcrowded parking lot, and didn't give it another thought. Why should I give it any thought?

After all, Google didn't do maps.

Chapter 7

SERIES B OR SERIES G?

"Do we have to take the money?" I asked, breathless as I tried to keep up. In our first few steps, I could usually tell how well Keyhole was doing by the pace John set. An uneventful day at the office, and we maintained an easy jog. A missed product milestone or a delayed push of a new imagery database, and we galloped. Throw in a kerfuffle with Brian McClendon over some user interface decision, and we'd be off at a full sprint.

Often John would stop by my cube near the end of the day and say, "Let's go for a run." I'd stop whatever I was doing and throw on my running clothes in the bathroom; then out the front door we would go. We would walk in silence along the sidewalk, John seemingly gathering his thoughts as we passed Microsoft's Silicon Valley campus on our right. A block and a half down the road, we would reach the trailhead, where La Avenida dead-ended. We would stretch out a bit and then hit the crushed-gravel trail, running north along the Stevens Creek, which split the massive 2,300-person NASA Ames Research Center across the creek on our right (where NASA's space-mission-controlling software was written) from a trailer park on our left (where you could buy an $800,000 mobile home).

John ran faster. "We have to take the money if we want to get there first. I mean, we are clearly on to something. Something with the opportunity to be big. Like ten times as big where we are today," he said. At this point, Keyhole had about fifty thousand paying subscribers. "The board wants us to go for it, and we're likely to have competitors gunning for us soon." John sprinted up the hill above the Shoreline Amphitheatre and around a bend. I lost him.

That spring of 2004, as the tech sector began to show signs of life after its three-year contraction, venture capitalists had woken up and once again began to open their wallets. As a company that had weathered the dot-com crash and had, step by step and year by year, transformed into a break-even enterprise, Keyhole was an attractive investment. Many companies up and down the 101 had thrown in the towel, but we were finally in control of our own destiny.

I hadn't been a part of the effort to raise capital: The $10 million Series B investment round would value our little mapping upstart at $30 million. Throughout Silicon Valley, a new season of innovation was being seeded with capital under the nebulous moniker Web 2.0. John knew the time was right. He hoped to use the new money for more data, more servers, more tech talent, and even a bit more marketing.

In February and March of 2004, John and Noah Doyle shopped the Keyhole investment opportunity along Sand Hill Road, a four-mile stretch that runs through the heart of Silicon Valley and up into the Santa Cruz Mountains, and is considered the venture capitalist version of Madison Avenue in the advertising world. John managed to secure three investment term sheets from different VCs, but it was Menlo Ventures, one of Silicon

Valley's oldest and most highly regarded VC firms, that offered the most agreeable investment terms. A deal was imminent.

John, sweaty from his fast sprint, waited for me at the top of the hill, overlooking the San Francisco Bay to the north. "Listen, Kilday, you understand that we have got to deliver on the numbers. You know I won't be able to protect your job or even *my* job if we don't hit our numbers."

"I know. I *know*, John. I got it!" I said, waving him off while I tried to catch my breath.

It was the second time that week he had hit me with this threat, and he was starting to freak me out. These afternoon runs were a zero-sum stress game for me: Every stress-reducing, dopamine-releasing mile was negated with an equal and opposite adrenaline-and-cortisol-inducing mile as John peppered me for status updates and marketing minutiae.

No detail was too small. "I thought you said the monthly newsletter would go out today?" "You told me two weeks ago that the new demo video would be out in ten days." "Didn't we agree to make that button blue?" "Why haven't you updated the pricing matrix to reflect the new Earthfusion price scheme?" "Can you see if you can get the cost of the new icon set down to under $2,000?" "I thought our last run of spec sheets came out too glossy. They felt cheap."

Mile after mile he would continue. John was aiming to change the world, but from my perspective, this Menlo money amounted to only a shit-ton more stress for this marketing guy.

Nevertheless, on Wednesday, April 21, three original copies of the Menlo Ventures term sheet were delivered to Keyhole. The document spelled out the terms that had been verbally discussed. Doug Carlisle and his partners were ready to complete the deal, and a meeting to sign the term sheet had been scheduled for April 26. The office was abuzz with excitement. In fact, I wasn't sure if much work was getting done. Games of pool and darts started earlier than usual that day. (We had a house rule of no pool before 4:30 P.M., but that day was an exception.)

But Monday came. And Monday went. With no announcement from John. In fact, John was holed up in a conference room with Brian, Michael, and Noah. As I walked to the kitchen or bathroom, I fleetingly glanced through the porthole window of the conference room door and saw them still gathered around the table. When I left work at 7:30 P.M., there were no signs that they were going to finish up any time soon. I, like many others, began to worry. Had something gone wrong with the deal?

When I arrived the next day, it looked like the four of them had never left, as they were already back in the conference room. Finally John emerged late that Tuesday afternoon. He looked relieved. I got the sense that something was up and commented on the marathon-like nature of the meeting. In response, John asked me if I wanted to grab a beer after work. We both drove to the Sports Page, parking next to each other in the lot, which was unusually crowded for a Tuesday. I was surprised that John was waiting for me in the parking lot because we often met inside the bar. As I got out of my car, John stood at the rear bumper of my car, blocking my path to the bar's front door. His expression was stern.

"Hey, listen, I need to tell you something."

Looking around, John continued: "You've got to promise not to tell anyone. Not even Shelley."

"Okay."

John stood there in the middle of the black asphalt parking lot. The sun was setting against the pristine blue sky. He paused as another group of strangers walked past us. A cool breeze pushed through the nearby stand of willow and palm trees. A raucous cheer erupted from the beach volleyball court adjacent to the bar. More cars streamed into the parking lot. John turned to me after triple-checking to ensure no one was within earshot.

"Google wants to buy us."

I stood there, my mouth open, I'm sure. John turned to walk toward the bar's entrance. I stood there. I wasn't sure if I'd heard what he had said. After a few steps, John paused and swung around. This time, a sly grin played across his face.

"What?" I asked, shaking my head.

"That's why we haven't signed the Menlo deal."

"That makes no fucking sense. *Goooooogle!?!*"

"Keep it down, Kilday," John said, ushering me toward the bar.

I started walking again, albeit slowly. Google? The same Google that had—that very day—finally announced the date of their IPO? It was the lead business story of the day around the world, a watershed moment in Silicon Valley. As the latest darling of the tech start-up community, the IPO valued Google at a whopping $27 billion (with a *b*!) valuation and was a signal to the world that, after four tumultuous years, Silicon Valley was back.

John was being particularly secretive because of who he had seen. Googlers. By the Prius-load. Streaming in to celebrate their big IPO announcement. Google, as it turned out, was also about four blocks from the Sports Page, and John was worried that he might run into someone who knew about the deal. While he would have preferred to wait to tell me, he surely didn't want me hearing of the news from someone else.

I fought to find a free table. I finally grabbed one along the wall as John bought us a pitcher of beer. It took a while because the bar was so crowded with Googlers. As I sat there, still stunned, I formulated a list of questions, but one was paramount in my mind: Why the hell would Google want to buy Keyhole? Google didn't do maps. It had no mapping product. Zero. Zilch. Nada.

John navigated through the knots of people and sat down at the table, beer spilling over the lip of the pitcher. He poured us both a beer. John explained that Google had become enamored with the Keyhole technology, but didn't give any specifics about the company's plans for us.

"We don't know yet, but we'll get more details soon. The company is putting together an offer," John quietly explained, not wanting to utter the word *Google* among its celebrating employees. "The other deal is being put on hold."

One Googler in a black fleece Google jacket roamed the bar, taking group snapshots with his high-end Nikon D70 of the young soon-to-be millionaires, many only six years removed from delivering their high school valedictorian speeches.

"The IPO is August 19. That's four months away," I said, recounting the day's news. "Do you think we can get the deal done before the IPO?"

"Oh, definitely," said John. "We'll get it done in a month or two—and *then* you can tell Shelley."

Finally I worked up the gumption to ask the question that I truly wanted answered.

"So, will I still have a job?"

"Yes," John said, "you will have a job."

I was surprised by John's confidence and formality in making this promise to me. It is common knowledge that marketing professionals are often the first to go after an acquisition. An acquirer buys the technology and the customer base, not the marketing team.

By the time I returned home to our rental in Palo Alto, Shelley was already putting twenty-month-old Isabel to bed. "No, Mommy read," Isabel said when she saw me (she was very advanced for her age).

"How was your day, honey?" Shelley asked.

"Oh, fine," I said, lingering a bit longer to watch them.

The next day, John and I went to lunch at the cafeteria at Shoreline Golf Links, a public course in Mountain View, and he filled me in on the rest of the story. Two days before getting the term sheets from Doug Carlisle at Menlo Ventures, John had received a call from Megan Smith, head of corporate development for Google. Megan completed her master's studies in mechanical engineering at MIT and was the founder of PlanetOut. She is an excitable fire starter, always smiling with a positive attitude. Was John available to come over to the Google campus and do a demo? She explained to John that Google cofounder Sergey Brin and other execs had apparently seen EarthViewer and were hoping to learn more. She did not then explain how that interest had first surfaced, an episode we learned about much later.

It went something like this: Two weeks earlier, a dozen executives met at Google for a product review of the company's photo-editing software, Picasa. Midway through the meeting, Sergey Brin, the brilliant Russian engineer, flip-flopped into the meeting fresh off the sand volleyball court at the center of the Googleplex. Opening up his laptop, he soon became distracted by something Google engineering head Jeff Huber, a good friend of Brian McClendon's, had sent him. The Picasa product manager was a few slides into his presentation, but began to notice that others were distracted by what was on Sergey's screen, too. Eric Schmidt stopped the meeting and asked Sergey if he cared to share what was so important. With that, Sergey jumped up, swiped the projector from the jilted product manager, and started projecting what was on his laptop: Keyhole EarthViewer.

The darkened conference room fell spellbound. Sergey conducted a free-form tour of the globe; the Picasa review was all but forgotten. Instead, like the Intrinsic Graphics presentations of four years earlier, the meeting was hijacked by EarthViewer, and the previously bored execs leaned in. They were mesmerized, several jumping up and down in their chairs, pleading for Sergey to type in *their* home addresses.

The commotion died down. Without a hint of business strategy to back him up, Sergey simply said, "We should buy this company."

The Google execs looked around the room at one another, and no one disagreed. Apparently, that was that.

After lunch, John and I walked over to the driving range to hit a bucket of balls. As we took turns watching each other hit shank shots, John continued. Megan Smith had opened the door with Keyhole, and the deal moved quickly. Michael, Brian, and John made the six-block drive to Google and demoed Keyhole on Monday afternoon for a group of execs. They met with Jeff Huber and several others, including Megan, Larry Page, Sergey

Brin, and Eric Schmidt. The demo had gone extremely well, and John discreetly let Megan know that Keyhole was close to taking a round of financing from Menlo Ventures.

The next day Megan called John and asked if she could come by the Keyhole offices. John described the meeting as a "rather obtuse, meandering conversation," with Megan wanting to "hear more about our plans" and "explore working together." John said that it was strange. Google, after all, had no mapping service.

"It's like we were about to get married, and suddenly the most attractive suitor on the planet came calling," John told me. Or at least he thought. Or maybe we were just being teased?

On Wednesday, the Menlo Ventures term sheet arrived. With the document in hand, all John needed to do was to sign on the dotted line. And if we took the money, it would set us off on a four-year path on our own. Without Google.

In a breach of early dating protocol, John picked up the phone that afternoon and called Megan and asked, "What's going on here?" He told Megan that he had received the term sheet from Menlo Ventures. "Is there any reason I shouldn't sign these documents?"

Megan showed John her cards: She told him that there was genuine interest at the highest levels of Google in acquiring Keyhole. To John, this prospect was incredibly exciting: The capital and technology resources of Google far outpaced any potential venture capital infusion. Google was a launching pad. A platform on top of which a service like Keyhole could infinitely grow.

But it was also somewhat curious. Google didn't do maps. What were its plans for Keyhole? Truthfully, neither John nor Megan knew the answer—and the answer to this question had to be shelved. John did know that he would need to give Menlo Ventures an answer; the closing meeting had been scheduled for the following Monday.

John told Megan, "You guys have to do something now or an acquisition is not likely going to be possible."

Megan understood. She asked John to give Google a week. John agreed to stall the Menlo Ventures deal close for that period of time.

The following Thursday, Google made a nonbinding offer to acquire Keyhole. John gathered his core deal team to assess the two options in the conference room.

For Brian, having confirmed the rumors of Google's hundreds of thousands of servers, the decision was an easy one: He was all in. Michael didn't hesitate either. He knew how exciting it would be for the entire Keyhole team to be at the center of the tech universe and how much more quickly Keyhole's vision could be achieved at Google. Noah, from a purely monetary perspective, knew that the Google offer was much better than anything Menlo Ventures could put together. He supported the deal, too.

But for John, questions lingered about the Google option. What would it mean for the Keyhole vision? What were Google's intentions for the service? Google didn't do maps, so what were its plans for Keyhole? Would we be acquired and simply have our technology folded into some Google service, the team broken up into pieces for easy assimilation? Or did Google truly believe in the vision of a fast, fluid, 3-D model of the entire planet? And how about the Keyhole employees? What were Google's intentions?

In fact, Michael and John were not interested in countering

the price of the acquisition. Google had offered $30 million for Keyhole. Rather, John was more interested in ensuring Google's commitment to the original vision of the idea—to deliver a high-resolution 3-D model of the entire planet.

They jointly agreed that John needed another meeting at Google to ask its founders more questions. On Tuesday, May 13, John and Michael met with Larry, Sergey, and Eric again to discuss Google's plans for Keyhole. They met in the founders' shared office. Dirty athletic clothes, hockey pads, and disassembled toys were strewn across the carpeted floor. Hockey sticks leaned against the walls. (Sergey was an avid roller-hockey player and organized highly competitive games that were frequently played in the Google parking lot.) Next door was Eric's adjoining office.

"Where do you see this concept of a 3-D model of our globe going?" John asked.

"I think this is something that could be core to Google," Larry said. "There would seem to be so many types of data that could be organized around a map and geography."

Eric added, "I promise you, we will get you more imagery data than your team could ever handle."

After the meeting, Megan pulled John aside near the snack bins on the second floor of Building 40. "Hey, I want to show you something," she said. "Something I'm really not supposed to show you." Up to this point, John, like many others, could only wonder if Google was generating the kind of revenue that had been speculated. Or was this just another dot-com with flaky financials? Did anyone ever click on those tiny little text ads? Because Google was a private company, no one outside the company knew.

Megan flipped open her laptop and showed John a spreadsheet of Google's financials for the past three years.

"Holy crap."

He had never imagined that a private company could produce those kinds of numbers or be that profitable. John was stunned.

With the full support of the Google founders and CEO, and a glimpse behind the curtain of Google's unprecedented financial success, the decision was easy.

John made the difficult call to Menlo Ventures. He did his best to be considerate, given the circumstances.

"We're not signing the documents. We're going in a different direction," he told Carlisle.

"Can we at least try to match the offer?" a stunned Carlisle asked.

John wasn't able to tell Menlo Ventures whom they were competing against; Google had entered a silent period leading up to the IPO. Instead, he could only say, "No. I'm sorry. You cannot, but we have an offer to be acquired."

Weeks later, John told me that Menlo Ventures and Carlisle had been supportive of Keyhole and the acquisition. It would have been easy for Menlo Ventures to make the negotiation difficult, considering where Keyhole was in the deal process, but Carlisle and Menlo Ventures chose to simply step aside and let John and Keyhole proceed with the acquisition.

Outside of Michael, Brian, and Noah (and soon me), John couldn't inform the Keyhole team of the seismic change in plans. Throughout May, negotiations with Google progressed rapidly, and by the end of the month, John was optimistic that the acquisition would be completed before the IPO. Being granted "pre-

IPO" shares would be an additional win for Keyhole employees and shareholders.

Then, at the eleventh hour, as the deal was set to be signed, disaster struck. It came in the form of a lawsuit. On May 28, Skyline Software, one of Keyhole's little-known competitors, sued for patent infringement. Filed in Boston, the suit claimed that Keyhole was violating Skyline's wide-ranging patent related to the remote landscape display and pilot training.

Of course, we had heard of Skyline: The Israeli-based outfit operated tangentially in the same space. That said, their service was limited to single cityscapes for flight training and other governmental uses, such as urban planning and military. They had relocated their headquarters to Washington, D.C., in order to pursue federal government and military contracts. I test-drove Skyline once and experienced frustration with its comparatively limited and slow service that forced you to switch between databases of various cities. For example, you didn't fly between Phoenix and Las Vegas; you had to switch between the two different models of the cities.

The claim was a broad reach, using a patent that lacked specificity as to what exactly it protected. Regardless of its merits (or lack thereof), the Skyline suit wound up pushing the Google acquisition of Keyhole out for several months. It was an excruciating wait for Keyhole and all its employees. The stress took a toll on John as he did his best to keep the day-to-day operations of the company moving forward—from adding new data to renewing customers and fixing bugs—all the while watching from the sidelines the biggest news in the business world during the summer of 2004: the Google IPO.

That summer, John and I continued our runs together after work. One early evening in late August, we ran up the hill, actually a landscaped former landfill that separates the Shoreline Amphitheatre from Building 41 on the twenty-six-acre Googleplex. Methane vents dotted the barren hillside to release gas from the decomposing landfill below our feet. The shimmering lights of Building 41 started to glow as the sun disappeared behind the Santa Cruz Mountains in the distance. Recently, more than one photojournalist with telephoto lenses had been shooed off the hill by Google security during the lead-up to the IPO. (Googlers in Building 41 had been instructed by the head of security to turn their monitors away from the expansive windows.)

The GOOG stock had debuted on August 19 at the lofty price of $85 a share. Larry's letter to the potential shareholders was unconventional: a call to arms that Google intended to make big bets, with a long-term vision, and that the company would not be overly concerned with short-term quarterly profits. Despite these warnings, the stock immediately shot up over $100 a share the day of the IPO. While Keyhole tried to resolve the Skyline lawsuit, every dollar that the stock went up was a dollar that our team missed out on. John and I lingered in silence on top of the hill, taking in the sunset over Building 41. Eventually, and reluctantly, we turned around and ran back to Keyhole.

RECALCULATING

The Google Years

Chapter 8

FEELING LUCKY

"We will bury you!" the testy Israeli yelled at John, pounding on the table at the law firm. Unbeknownst to many on the team, including me, John had flown to Boston to sit down with a mediator and Skyline in hopes of negotiating a settlement in person. John opened the conversation with what he considered to be a reasonable offer. Let's just say that Skyline did not agree. Since the Google IPO in August, there had been little progress with the lawsuit. While Google had been supportive, they had also placed the onus on John and Keyhole. This was not Google's fight to fight: It was up to John to resolve. After the trip to Boston, John resigned himself to the idea of a protracted legal battle in court.

John was suspicious of the uncanny timing of the lawsuit. Had Skyline caught wind of the deal when he and Dede notified shareholders of the offer to be acquired? Skyline's negotiation made this explanation plausible.

Megan Smith appeared at our offices frequently to meet with John and to keep some form of momentum going. Always smiling and positive, Megan filled in those weeks of waiting for Keyhole with an attitude of resilience in the face of inertia. Like John, she was going to get the deal done—whatever the obstacle.

John tried to remain optimistic too, but it was a frustrating waiting game for him and everyone around him. I stopped asking him for updates. Most of the company, and their significant others, were desperate for any signs of movement on the deal.

Finally, David Drummond—head of corporate development at Google and Megan Smith's boss—stepped in. It is possible that Larry and Sergey had pressured him. In late September, he told Megan: "Lawsuit or no lawsuit, let's get the deal done."

At last, in late September, the wheels on the acquisition deal started moving again even though the Skyline lawsuit remained unresolved. To cover its legal cost exposure, Google added a clawback clause to the Keyhole acquisition contract, meaning that legal fees incurred by Google would be subtracted from the purchase price and held in escrow until the Skyline matter was settled.

On a Friday afternoon, John was finally ready to tell the whole team. It was the worst-kept secret in our small company's history, but there were still a handful of people that didn't know about the Google deal. Unlike most company TGIFs, John let everyone know that attendance at this Friday's gathering was mandatory. Mark Aubin fired up the grill to cook sausages. John requested that everyone return with their plates of food to the conference room. All twenty-nine of us. In a room with ten chairs. I wasn't the only one to take note when John Rohlf sat on the floor along the back wall. It might have been a first. John Rohlf at a company-wide meeting.

"I know that some of you—no, several of you—already know this news, but I'm letting everyone in the company know the news today officially," John said with a broad smile. "We have agreed to be acquired by a little company down the street." The entire team erupted in applause. Wayne Thai shouted, "By who? Microsoft?" He pointed across the street.

"Nope! *Google!*" John replied, a sly smile on his face. Another round of cheers and high fives swept through the conference room.

John informed everyone that we would need to interview for our jobs at Google. He assured us that this was largely a formality, really more about job "leveling" (read: deciding what level we would come in at). As a team, we would not be subjected to the standard Google interview process, which was known to weed out all but the highest achievers from the top universities.

John had required that the entire Keyhole team be hired as a prerequisite of any negotiation talks with Google. In the tech acquisition world, this is unheard of. More commonly, huge swaths of teams are not hired, being paid for their equity in the company but not given jobs. When Megan Smith first told John of Google's interest, he said, "That's great, but to be clear, you would need to commit to hiring the entire team. If this isn't a possibility, we shouldn't waste each other's time." It was not the first time—and would not be the last—that I would come to understand that John Hanke was the most loyal boss and friend one could hope for. There were twenty-nine of us at Keyhole at the time, including Mark Aubin, who had tried to quit recently. John had only told him: "Don't quit just yet. I can't tell you why, but don't quit."

On a breezy Mountain View summer afternoon, I walked eight blocks to Building 41 on the Google campus. I should have been nervous. But after walking along the eucalyptus-tree-lined streets leading up to Google's gleaming new campus, I felt refreshed and relaxed.

My HR escort badged us through the double glass doors into one of their well-appointed conference rooms on the ground floor. During my interview, I met with four Googlers: David Krane (head of PR), Doug Edwards (director of marketing), Debbie Jaffe (my soon-to-be boss, head of consumer marketing), and Christopher Escher (lead designer). I'm a perennial note-taker, and when I opened up to a blank sheet as I sat down with David, I joked with him that I needed Google to scan all my old spiral notebooks so that I would be able to search through them.

"Oh, you'll be able to do that one day. We're working on scanning books in libraries using book-scanning robots," he said. I laughed until I realized he wasn't joking.

During our meeting, Doug and Christopher were more interested in learning what Keyhole would be doing at Google, asking me lots of questions. It was all cordial and informal, though Debbie Jaffe did treat it as a real interview. Almost like a walking flow chart, Debbie took me through the way marketing worked at Google in great detail.

If *anyone* at Google knew why Keyhole was being acquired, it had certainly not been shared with David, Doug, Debbie, or Christopher. They knew even less than me!

At Keyhole, it was a nervous week of interviews. Every thirty minutes or so someone stepped through the door of our office and reported to John about their interview. He always asked the same series of questions: "Who did you meet with?" "What did they ask you?" "How do you think it went?" Each interviewee returned to our offices, relieved.

On August 25, Megan Smith and Sukhinder Singh visited the Keyhole office to present twenty-nine offer letters. I was called into the conference room, congratulated, and presented my letter by Megan.

All twenty-nine Keyhole employees were hired. Just as John had promised.

That day I left early for Isabel's second birthday: an ice cream party on our tiny patio in Menlo Park. Shelley and I sat on the sun-splashed flagstone steps, watching Isabel and her two-year-old friends play. I had an ice cream cone in my hand and the offer letter in a Google envelope in the back pocket of my jeans. Finally I felt like I could enjoy a moment like this—with fewer worries about Keyhole and our future.

Later that week, Bret Taylor, an ambitious young product manager at Google, met with me at Keyhole's offices. He had been assigned to a new mapping project and was working closely with Danish brothers Lars and Jens Rasmussen of Where2Tech, a small team acqui-hired by Google three months earlier. (At the time of the purchase, the four-person team was not incorporated and was based partly in Demark, Australia, and the United States.) That day John was at a different meeting over at Building 41, so I presented Bret with a PowerPoint of Keyhole's product road map and how we would handle the marketing transition of Keyhole into Google. We would drop EarthViewer product prices across the board by 50 percent, but that would be the only material change at acquisition. For the immediate future, the product would still be called Keyhole EarthViewer and appended with the words *powered by Google*. Once we were fully integrated with the company, we would "rebrand" the product with a Google name. We agreed that was likely months away.

At the end of the meeting, Bret asked me, "Do you think that you'll be the PM [product manager] or the PMM [product marketing manager] after the acquisition?" Knowing only a little about the difference between the two roles at Google, I tentatively answered, "Well, I'm both right now. I manage the product features and I manage how the product is marketed. So I think I'll do both roles."

It was true: At Keyhole, I had performed both positions. I helped to build the actual product, including prioritizing features, sketching out interface schematics, hiring user interface (UI) designers, and managing the product backlog. In addition, I helped market the product, which included creating our websites and sales literature, pricing the products, executing trade shows, writing taglines, and creating demo movies.

Bret looked at me with mild amusement and laughed. "Well, it's not my place to tell you that you can't do both," he said, "but I don't think it will be possible. I mean, you can try. But eventually you'll figure out on your own that being both a product manager and a marketing manager isn't possible at Google."

Over a beer at the Sports Page that evening, John was interested to hear about my meeting with Bret. "What else did he say?" John pressed. As always, John was thinking a few moves ahead in the chess game, a game that I didn't even realize was being played.

"Be careful with Bret," John cautioned me. "We don't know yet where we will all plug in at Google. There's a top executive who has an interest in maps, and Bret is one of her up-and-coming stars. I don't want us all reporting into her through him."

That was crazy talk to me. As capable and sharp as he was, Bret was *young.* Like twenty-four years old young. Sure, he was well spoken, the state champion of such and such, spoke multiple languages, and wrote killer code. But I couldn't imagine having all of Keyhole report into Bret.

During that same week, another meeting at Google in Building 40 focused on privacy concerns. Representatives from a cross-functional working group were in attendance to learn more about privacy and aerial imagery, and to talk about how the investment from In-Q-Tel might be perceived.

Various legal precedents related to the capture of aerial imagery made it clear that Keyhole was operating on safe legal ground. But that was of less importance to Google than was the public's perception. It was made clear to us that the trust in the Google brand among users was of ultimate importance. We should not screw that up.

At all of these introductory meetings, I was amazed at how little anyone talked about pricing and sales and revenue forecasts. It was a foreign concept to many at Google that we charged for our product. In fact, generating revenue seemed to be of little importance among its various teams. Instead, the Googlers were focused on ambitions like user delight and changing the world with technology. Making money was rarely, if ever, brought up.

At another preacquisition meeting in Building 41, we discussed getting our website, keyhole.com, ready to be relaunched as part of Google. It was going to be revised with the messaging about the acquisition ("Keyhole's Feeling Lucky"), new pricing, and updated information about privacy and sources of data.

Patricia Wahl, Keyhole's hyperorganized webmaster and documentation lead, came with me to meet with Karen White, Google's webmaster. After reviewing the structure of our website and marveling at the way that Patricia had named and organized files, Karen stopped the meeting, turned to Patricia, and said, "Wow! Do you want a job on my team?"

Patricia and I laughed. Karen said, "No, I'm serious." She wasn't kidding. It was my first taste of the new opportunities available to the Keyhole team within Google.

There was one playful design element that I wanted to include on the new Keyhole website "powered by Google." In addition to the Google logo, the company often used an icon of five balls of varying colors as a sort of spacer or whimsical visual breather. Often, this design graphic was placed on the bottom of a web page or a piece of marketing literature.

I wanted to alter those balls and use them on the Keyhole site, subbing in a tiny blue marble Earth for the blue ball. Karen liked the idea, but said that I would need to check with Marissa Mayer if this small design modification would be okay. I had no idea who this Marissa was. Over the past two weeks, I had met so many people and heard so many names. In passing, I had seen a fashionably dressed young woman with bright blond hair in a glass-walled conference room with Debbie Jaffe, and had mentally filed her away as a Googler who worked *for* Debbie. Maybe that was Marissa?

Two weeks passed and much prep work was completed: updated pricing, approved press releases, embargoed interviews, FAQs, quotes from executives, new help center content. Still, I hadn't received approval of the blue Earth ball idea from Marissa. It was a minor detail, but I was attached to it. As the announcement neared, I kept on Karen. Still, she hadn't received any approval from this Marissa either.

I emailed Debbie, going over Marissa's head (or so I thought), for approval. Debbie also liked the idea, but reiterated that it was something Marissa needed to see. I took that as Debbie's desire to empower the people on her team. Frustrated by numerous unanswered emails from this junior marketing person, I decided to call Marissa's cell phone and get this taken care of once and for all.

My call went straight to voice mail. I left a terse, unapologetic, and demanding message. All I needed was a quick email approval. Karen, Debbie, and others were on board with it—and I was just asking for her thumbs-up. "I've sent six emails and have gotten zero response from you. I really don't understand what the delay is. I need to hear something from you today or I'll have to take this up the ladder with Debbie and John Hanke." I was annoyed that this junior, first-job-out-of-college young woman was blocking my creative idea—and my tone certainly let her know that I found this unacceptable.

Noah Doyle sat in the cube next to me and popped up like a chipmunk when he heard my message, trying to control his laughter. "The thing is. That Marissa. Well, the thing is, she's kind of a big deal."

Noah went on to tell me that yes, Marissa was, in fact, the young woman I had seen around with Debbie Jaffe, but as

Google employee No. 20, and its first female software engineer, she ran *all* of Google search. I had just dressed down probably the most powerful person in the company. Noah added, "Yeah, she's arguably the most powerful woman in the entire technology industry. I think more than half of Google reports up to her."

Dede had heard the entire exchange. She was laughing, too. Come to think of it, she was laughing uncontrollably. If I was going to have to work with Marissa—and chances were I would—I was off to a smashing start. Eventually I figured out that I needed to attend Marissa's user interface weekly team meeting and ask for her approval in person. After I waited forty-five minutes for my turn on the agenda, it took her all of one minute to look at the slide I was projecting in front of her team of designers to say simply, "Oh, that's cute. You should definitely do that." It was approved.

As the acquisition deal neared the finish line, John decided that he would assemble a small Keyhole team of eight for a planning retreat off-site at the Asilomar Resort on the Monterey Peninsula. The evening we arrived, John led the team on a long walk that traversed the coastline. He talked about Keyhole's chance to start over, but also how we were heading into uncharted waters. During our years together at Keyhole, we might have been a part of something great, but now we were heading toward something we didn't know. As a result, we needed a plan, because if we didn't have one in place, there were plenty of Googlers to tell us what to do—and our team would be splintered apart.

Hypersensitive to any information leaking out, we didn't speak or write *Google*, especially not on any of the whiteboards or giant Post-it notes scattered about our breakout room.

We talked a lot those two days about a company we had long since decided to not compete against: MapQuest. Noah had assembled some eye-opening market data about its purchase by AOL, the number of monthly visitors, and a back-of-the-envelope calculation of the dollars per user per month MapQuest was likely earning. With two thirds of all Internet mapping happening on MapQuest at the time, the company was the global leader. Up until then, Keyhole had purposely stayed away from the consumer web mapping space. But now, with the resources of Google behind us, we began to wonder: Could we consider a web mapping service to rival MapQuest?

We also talked a lot about the Where2Tech four-person team. The initial demos of the technology looked quite impressive. The advanced JavaScript methods of prerendering map tiles in a browser being worked on by Lars and Jens Rasmussen sped up the user experience, and Bret Taylor and Jim Morris, another top Google engineer, had jumped in to lead the effort to bring that technology onto Google's platform.

We wondered if the fast panning of map tiles in a web browser could be combined with the aerial and satellite imagery of Keyhole. And what if that could also be combined with the rapid, relevant searching capabilities of Google? It seemed like an exciting concept.

It was also a concept that was sad to consider. At Asilomar, we started to come to the realization that the best thing we could do for Google—the biggest impact Keyhole could have for our new owners and users—might mean the end of the Keyhole team. It would mean splitting our energy and our team in two. One Keyhole team would work on building a fully Google-fied version of our flagship EarthViewer product. I, and many others, would remain on that team. A second Keyhole team—led by Chikai—would focus on helping the fledgling Google mapping project. John would oversee both. All Keyhole satellite and aerial imagery would be imported into a new web-based product that would soon be launched. This half of the team would pivot away from our 3-D roots. John, Daniel, and Chikai would have a mountain of work in front of them, redoing the database and infrastructure to integrate aerial and satellite imagery data into a new web-based mapping product.

Of course, this meant the Keyhole team would be inserting itself into a project that was already under way at Google. And this newly formed Google mapping team was functioning quite well without us. Nonetheless, without being asked, the Keyhole team decided to put Google's web-based mapping efforts first. But did that team want our help? There were four individuals on their team, and twenty-nine on Keyhole's. This led to the next obvious question: Who would *lead* this effort?

I was starting to get an idea of the border dispute mounting for control of Google's unreleased mapping products. But John already understood this. It was the reason that he had cautioned me about Bret Taylor. He knew that Bret was ambitious and confident. Also, he knew whom Bret worked for and where his allegiances lay: the person running all of Google search.

"Wait," I interrupted John. "Are you saying that Bret reports to . . . to . . . *Marissa*?"

"Yeah, he does. And Google higher-ups might interpret a map search as just a different flavor of search," John explained. "And you know, of course, that Marissa runs all of search. I think she wants to own maps, too." I slunk lower in my chair. Noah laughed.

Later that week, on Friday, October 15, the acquisition deal was supposedly done, but the countersigned documents had yet to be delivered to Keyhole. The word from Megan Smith was that she was trying to track down David Drummond for his signature. Apparently, he was flying in from being out of town and she didn't know when he would be back in the office. We went through the motions as if it were a normal workday; after all, there was plenty of work to do while we waited for the final contract to be hand-delivered.

At three o'clock, a delivery truck wheeled in a giant rack of yellow Google blade servers with *forty times* the capacity of our current server backend. Google engineers privy to the deal were already nervous about the server load that the acquisition announcement would likely generate. Within an hour, all pretense of real work getting done was gone. Even John loitered around the office impatiently. We played pool, darts, shot baskets in the parking lot. Finally, at six o'clock, the official original of the acquisition of Keyhole by Google Inc.—signed, executed, and dated 10/15/04—was delivered to John. He looked relieved to finally have the document in his hands.

John uncorked a bottle of inexpensive champagne that had been sitting on ice in the lobby all afternoon and said a few nice words to the dozen or so Keyholers—now officially Google employees—who had stuck around the office. He told us to report to Building 41 on the Google campus at nine o'clock on Monday morning for orientation. He also reminded us that while the deal was finalized, it had been agreed that there would be at least two weeks between the deal close and the public announcement of the acquisition.

That evening Keyhole had its own celebration; I convinced John to loosen the purse strings one last time and rent out a private room at a restaurant called NOLA, a festive New Orleans–themed restaurant in downtown Palo Alto. (The $1,200 room fee was the very last check written out of the Keyhole business account.) I set up a projector that showed Keyhole EarthViewer, piloting around the globe to several of my favorite places of interest, jumping from one KML placemark to the next. The Red Sox played on the muted TV, but I paid little attention to the play-off game, as they were losing to the Yankees once again, likely going down three games to none in their best-of-seven series.

Several Google executives, including Jeff Huber and Megan Smith, were invited to our celebration. There were many toasts given that night, including those by John, Michael, and Brian. Lenette pressed me to give a toast, literally pushing me to the front of the room. I stood up on a chair and told about an incident from earlier in 2004.

The story goes like this: During the spring of that year, John and I jumped into his Subaru in the Keyhole parking lot to go to lunch. As was often the case, John was a bit stressed and in a hurry. As he whipped around the corner of our building, Chikai had shot out the office side door, also a bit too fast. He stopped with a jolt, and John slammed on his brakes. Startled, they both caught their breath.

I said to John, "Man, of all the people to run over, Chikai would be the worst." This was probably true: Chikai directed all data importing and mosaicking and publishing.

As we continued to lunch, I kept the game going with John. "Well, who *could* you run over?" I started down the list. "Patricia?" Oh, no, we couldn't possibly run the website without Patricia. "Aubin?" Oh no, what he is doing on the data processing tools is absolutely critical. "Daniel?" No way, all data deals and partnerships. "Lenette?" No, she's handling all our operations.

By the end of lunch, we had exhausted the list. *All twenty-nine.* It was true. All twenty-nine had been critical to the success of the business. It was impossible for John to name one individual that we could have done without. It was an amazing feeling—and one shared by everyone at Keyhole. It is a rare and wonderful realization. One that I will always remember.

As a team, with a bit of luck and a lot of hard work, Keyhole survived. We created a new way of looking at the world with our new map. And now, in October of 2004, with Google's bold backing, this powerful map was about to be let loose.

Chapter 9

TABLE FOR 33

On Monday morning, the Keyhole team reported for work at nine o'clock at the whimsical Googleplex. It was a typical Northern California morning; the hazy clouds hadn't quite burned off. Security told us to park in the designated visitors' spots until we received our Google badges. Meantime, countless Googlers arrived in the parking lot and streamed into the multiple buildings of the large campus. The Keyhole team—and about six other new hires starting at Google that day—assembled in the lava-lamp-filled lobby of Building 41. A scrolling list of real-time Google searches in multiple languages was projected onto the large wall behind two young receptionists. (We learned that a special filter was used to scrub the list to ensure nothing offensive appeared on the lobby wall.) John and Megan Smith waited for the entire team to arrive. We helped ourselves to the assorted jars of candy, chips, nuts, and high-end juice drinks. Our first stop in Building 41 was to pick up our official badges. Afterward, we stopped at one of the several microkitchens to refill on snacks again and make ourselves lattes.

"Here we keep the free Google T-shirts," Megan told us. "We ask that everyone be Googley and only take two or three your first day." And here's the gym. And here's the lap pool. And there's the beach volleyball court with Sergey on it. And there's the game of ultimate Frisbee going on. Here's the bike you can jump on to cut across campus. Or the electric scooter. Or the Segway. Here's where you catch the free shuttle bus that travels throughout the Bay Area. Note: There's a barista and juice bar at the depot for your commute. Yes, of course, the buses have Wi-Fi. There's the massage room for Building 41. Here, have some M&M's.

I'd say it took about eleven minutes to forget all about 94 La Avenida, with its Craiglist pool table, the leaking roof, and refrigerator full of brown-bag lunches and assorted half-eaten Costco dips. We had forgotten about all of these things until Dede brought a hazmat team from Google to clear out our old office three months later.

Next: Building 41's Techstop. Every building of every Google office features one of these stores with its fanciful stop sign. The store is staffed by two or three network administrators or IT support specialists whose sole role is to ensure Googlers have every desired piece of equipment functioning at peak performance. At Building 41's Techstop, we were invited to outfit ourselves with whatever tech gear we wanted or needed. Imagine a mini Apple store with no requirement to pay. Everyone grabbed notebook computer cases, chargers, software, and Wi-Fi routers that had been flashed with Google security software to facilitate connection to Google servers from home. (Google pays for its employees' Internet access at home, but requires use of a router with special security software.)

I noticed that the router being handed out, a blue Linksys brand model, was the same as one I had recently bought for myself. "Hey, I've got that same router at home. Can I just bring that one in and have you update the firmware with the Google security software?" I asked. I hoped to save the company $55. The Techstop staffer peered up at me from behind his two thirty-inch monitors and glanced at his counterpart, who seemed more than amused by my question. "Nah, just go grab a new one." *This* would take a long time to get used to.

We walked through a sunlit breezeway that connected Buildings 40 and 41. John suggested that we split from the team for a short walk because he had something that he wanted to discuss with me. "We need to decide on your title. Your job responsibilities are not going to change, but Google wants us to choose one title: product manager or product marketing manager." He too recognized that I had been both at Keyhole. He asked me my opinion on the matter; I didn't fit into the tidy job title of product manager at Google. Namely, I did not have a computer science degree. While that was not typically a requirement at the other high-tech companies, it was at the engineering-driven Google. That day, John and I decided that my title would become the imperfect product marketing manager, though my responsibilities were to be the same as they had been at Keyhole.

I was Google employee No. 2,488. (Today, there are 73,992 employees.)

Later that afternoon, John let us know the Keyhole team was scheduled for a meeting that even Megan Smith was surprised by. It was billed as an official "Welcome to Google" with Larry Page and Sergey Brin. Not knowing any better, I assumed that their appearance was standard for any new acquisition. Apparently, this was not the case.

"You guys are getting to meet with Larry *and* Sergey?" other

Googlers asked us repeatedly throughout the morning. It was worth more than a bit of Google street cred; various execs were taking notice of Larry's and Sergey's keen interest in Keyhole. They too were looking for signs and direction about what the heck Google was going to do with this small mapping team.

Six years earlier, Larry Page and Sergey Brin had been fellow Stanford PhD students who, despite not really liking each other at first, decided to work together on various projects. One of these projects was called BackRub, which examined the mathematical principles of the Internet. It was one of the five or six projects that they were working on. In order to put together a proof of their concept for BackRub, they downloaded the entire Internet (twenty-four million web pages at the time). Once downloaded, they ran a new software algorithm they had developed to filter and rank that database of websites, applying a new way of understanding how all those web pages were interrelated. The new BackRub algorithm determined which set of web pages to return to a user for a given search term.

This is how Larry and Sergey's innovation set itself apart from existing search engines: When ranking pages, they did not rank according to the number of times a keyword appeared on a given web page. Instead, they ranked web pages according to how many *other* web pages linked to it. It's sort of the equivalent of saying, "It's not important what *you* say about you; it's more important what *others* say about you."

For example, let's say you run a website selling Brazilian teak patio furniture. If someone went to Yahoo! and put in a search for "Brazilian teak patio furniture," Yahoo! returned a list of web pages that was prioritized by how many times the words *Brazil-*

ian, teak, patio, and *furniture* appeared on a given website. You moved higher up in Yahoo!'s list if you added multiple mentions of these words on your site. The end result: Users clicked on sites with repeated keywords, but not necessarily the sites of the highest quality.

In contrast, Larry and Sergey's algorithm took this approach: The number of times *you* use those words on your site is almost irrelevant. What is important is the number of times that *other* websites about Brazil and teak patio furniture link to your website. Other sites won't link to your site if they deem it low quality. They will link to your site only if you truly have good content.

To be clear, the BackRub concept didn't introduce a whole new way of ordering results. Truthfully, it was a creative interpretation of the way that respected science journals, like *Science* and *Nature,* had been ranking the relative importance of research papers for over a hundred years. In other words, by the number of *other* papers that referenced a research paper. An article with a high number of "incoming citations" from other articles or journals is considered to be of higher importance, especially if those citations are from other prestigious journals. I'd venture a guess that Larry Page would have been exposed to this concept by his father, Dr. Victor Page, a computer science professor at Michigan State University. He was considered a pioneer in the artificial-intelligence field.

Larry called his new algorithm PageRank (a clever double entendre), and they renamed their BackRub project Google, which was how Larry thought the obscure math measurement of a googol (one followed by a hundred zeros) was spelled.

By our first day in October 2004, Google had already exploded as a full-on cultural phenomenon. Larry and Sergey had become international tech celebrities, though it was clear from the get-go they had little interest in fame or material trappings. At the time of the IPO, when the stock price debuted at $85, they were both thirty years old and worth $6.7 billion. Larry still lived in a small Palo Alto apartment, and Sergey had recently bought his first car (an electric EV1 manufactured by General Motors).

Jonathan Rosenberg, the VP of Google's product strategy, met with us first, before Larry and Sergey joined the meeting. Confident and self-assured, Jonathan was always ready with a joke. Brian and John knew Rosenberg well: Brian had worked with Jonathan at Excite@Home in the mid-1990s, and when Keyhole had gone looking for a head of product and marketing in 2000 (the job I eventually took), Brian had recommended Jonathan for the position.

John Hanke and Jonathan met for coffee in Mountain View to talk about the Keyhole marketing job and mutually agreed that it wasn't a great fit. Instead, Jonathan took a job as head of product management for another Silicon Valley start-up: Google.

I was anxious to hear Google's plan for Keyhole. "A year from now," I asked Jonathan, "thinking about what success looks like, would it be better if Keyhole generated ten million dollars in revenue or had ten million users?"

It was an outrageous question, but I was throwing out big numbers to elicit some directional guidance about what we should focus on—making money or acquiring users. Our earn-out milestones, the goals that were established as part of the acquisition deal, had set a bar of 500,000 users for Keyhole. I was suggesting a goal of twenty times our earn-out target.

"Ask Larry and Sergey," he only said. "My bet is that they'd prefer the ten million users, but you should ask them that question."

The meeting was held in a college-like seminar room designed for various Google training programs in a remote building on the outskirts of the campus. Sergey Rollerbladed into the room, awkwardly maneuvering down the short flight of stairs to the front. Larry followed closely by him, smiling at Sergey's entrance. (Clearly, he had seen this stunt before.) John personally greeted the pair, shaking their hands, while the rest of us remained seated. We went around the room, with John introducing each one of us. Larry and Sergey were complimentary of the work we had done in launching Keyhole. It was clear from the beginning that both were interested only in the technology and math behind the model of creating a global mapping database.

"What percent of the globe do you have covered at one-meter resolution?" "From what source?" "At what resolution?" "Tell us about the satellites." "Who makes them?" "How much do they cost to launch?" "Are they geosynchronous?" "What are the size of the sensors?" "How large is each file?" "How many servers do you currently have?" "How fast do the satellites travel?"

Larry and Sergey debated each other about how long a satellite would take to capture the entire globe, accounting for darkness, cloud cover, landmass percentages, the size of each photo, the speed of the satellite, the rotation of the Earth, gravitational pull, fuel, weather, and other factors.

"How big would the database be if you covered the entire planet in one-meter resolution detail?" Sergey finally asked John.

I was certain that Sergey was joking. It was the equivalent of turning to Michelangelo after four years of work on the Sistine Chapel and saying, "Cool. How about painting the rest of Italy?"

John turned and looked up at Michael Jones, who was sitting next to me. "Michael, do you want to take that one?" "It would be approximately one petabyte," Michael responded without pause. It was the first time I had ever heard the word *petabyte,* which is one million gigabytes.

"I don't think that's right," Sergey countered with his own calculations. "It's five hundred terabytes."

We were almost out of time. Sergey continued to question Michael's math. He wasn't convinced. After a few more minutes of debate, Michael said with a degree of finality, "Trust me, Sergey, it's one petabyte. I'll come over and walk you through it in your office."

The pair started to leave the room. "Larry, Sergey, ten million dollars or ten million users, what would you prefer?" I blurted out.

They turned toward me, confused. "I do not understand your question," Larry responded flatly.

"If you think about what success looks like for us, for the Keyhole team, say, a year from now, would you prefer we add ten million users or make ten million dollars?"

Larry and Sergey looked at each other for a moment as if they were silently deciding who should take my question. With his trademark toothy grin, Larry said, "I think you guys should be thinking much bigger than that." He stared at John for emphasis. Silence filled the room.

And then Larry and Sergey left the seminar room, Sergey still buckled in his Rollerblades.

After the meeting, John led the Keyhole team back to Building 41. We started to unpack our moving boxes and decorate our new digs. There was palpable excitement as everyone settled into the brightly colored cubes, each space outfitted with a Herman

Miller chair and dual thirty-inch monitors. I found a ladder and started to hang the Keyhole and Google intersection street signs I had custom-ordered, but no sooner had I started than two members of the Google facilities team emerged. They were armed with twine, zip ties, measuring tape, and toolboxes. When they were done, the workers asked me to review the signs to ensure they were to my liking: Was the height right, the location, the angle, the way the twine was tied? Back at my desk, I received an email from the Google facilities issue tracker system, asking me to rate the interaction.

An ergonomics consultant made the rounds of Building 41, scientifically calibrating everyone's desk, chair, and monitor height to the optimal positions. We were given coupons for free massages from the on-site masseuses in case our setups weren't perfect. I moved into an office with Daniel Lederman, not knowing the protocol that only directors at Google were assigned offices. I had been director of marketing at Keyhole, but my level at the much larger Google dropped down a rung. I was a product marketing manager, but my office was situated right next to John. We enjoyed an uninterrupted view of the Shoreline Amphitheatre. We were now on the inside, looking out at the barren hill that John and I had often run up.

Brian and Michael moved into an office around the corner, anchoring our new section on the ground floor of Building 41. It was an ironic twist for Michael and Brian: Building 41 of the Googleplex had been a part of the Silicon Graphics campus some ten years prior. The same building—where Brian, Michael, Phil, Chikai, and Mark had worked on $50 million flight simulators—was at the center of this new Google mapping project.

The rest of the Keyhole team filled in a large section of purple-walled cubicles (also inherited from SGI). This area included four additional cubes for Google's upstart mapping project team, headed by Bret Taylor. While no one knew exactly what we were supposed to work on or how to organize ourselves, we did know this: We were all sitting together.

It was an uncomfortable group dynamic, when I think back on it. Imagine a hot new restaurant that has just opened, rumored to have the finest cuisine in town. It is empty except for four people sitting at a table together, enjoying their lunch. Then picture a much larger group of twenty-nine people bursting into the restaurant. They all know one another. In fact, they are long-time friends. They are intoxicated with the thrill and anticipation of this new restaurant. And now a new table is set: a table for thirty-three.

The four that were already seated, perfectly content (thank you very much), are now told to stand up, move over, and join the twenty-nine. They aren't given much of a choice in the matter. Now everyone is sitting together. There is no menu, no understanding of what to do, nor any direction about who is supposed to put the order in. But there is, apparently, an open tab.

Bret Taylor and Jim Norris had been at Google the longest. They had been reassigned from a team working on Google's efforts to improve answers to searches that included a location element. Bret and Jim had recently been paired with new hires Jens and Lars Rasmussen with the objective of reworking some initial Google mapping prototypes. Bret was the natural leader of that small team. Jim was a no-nonsense, quiet workhorse with a master's in computer science from Stanford, paying little attention to the inherent politics of a particular project. If there were two people at the ends of the table, John Hanke sat at one end and Bret Taylor at the other.

Those early Google mapping prototypes were engineered by Where2Tech. Its path to Google had been relatively short: The team had been working together for less than a year, but had already experienced their share of twists and turns and detours. In the fall of 2003, Jens and Lars Rasmussen had found themselves newly unemployed from a company called Digital Fountain and living in Sydney, Australia. Jens, part artist and part engineer, began tinkering with some ideas he had first pursued while working for a Yellow Pages publisher in Denmark in the mid-1990s. Soon Lars joined him in exploring these ideas, and they holed up in the bedroom of a friend's apartment, Noel Gordon. With a fourth friend, Stephen Ma, they began developing a mapping software application.

I say mapping *software application*—not mapping website—intentionally here. Where2Tech wanted to compete with Microsoft Streets & Trips, the leading CD-ROM-based mapping application of the day. The idea was a bit like Keyhole in that the user would install software on his computer, but most of the data would be served over the Internet from centralized databases. That said, Where2Tech didn't have any data yet.

Like Keyhole, they purposefully stayed away from building a browser-based mapping website because of MapQuest. And the security risk to data scraping (malicious software scripts that automatically download, or "scrape," a website's data) made it nearly impossible to pursue web mapping because providers of mapping data—like road network and business listing databases—refused to give Jens and Lars anything more than a few sample data sets. It was the same reason J. R. Robertson wouldn't allow Keyhole to build a mapping website: the fear that web-based mapping sites made it too easy for software developers to write scripts that would automatically steal all of their valuable data.

By early 2004, Where2Tech created a working prototype of its downloadable application. Without any funding, business model, or, for that matter, relationships with data providers, its prospects were limited. In March, Jens moved back to Denmark to pursue other work. In the meantime, Lars traveled to Silicon Valley with hopes of using their demo application to secure venture funding for Where2Tech. It was a hard sell. According to just about everyone in Silicon Valley, consumer mapping services were done. MapQuest owned the market. As John Hanke told me four years earlier, trying to go head-to-head with MapQuest for the consumer mapping market would be an expensive trip down a long, winding, and potentially dead-end street.

Surprisingly, though, Lars had a nibble, no actually a bite, from a venture capitalist purportedly willing to invest in Where2Tech. The term sheet was sent over for review, and Jens flew back from Denmark to Silicon Valley. A meeting was scheduled at the offices of the venture capital firm to close the deal. Like the Keyhole venture capital deal, the meeting never happened. This time, however, it was the venture capital partners who canceled the meeting at the last minute. They were scared out of the risky investment in Where2Tech by a seemingly minor market move by the distant number two player in consumer web mapping: Yahoo!

Yahoo! had made a small, relatively unimportant change to its Yahoo! Maps service on April 13, one day before the Where2Tech funding meeting. Yahoo!'s travel property—Yahoo! Travel—had long shown maps with its business listing results. That Monday, Yahoo! had provided these maps, and that functionality, front and center on Yahoo! Maps. It was a minor change, but it was an ill-timed signal for Where2Tech and the skittish investors that Yahoo! aimed to invest heavily in its mapping products. The funding meeting was canceled, and Where2Tech was still broke. Jens flew home to Denmark, and Stephen Ma and Noel Gordon stayed in Sydney.

Lars continued to network throughout Silicon Valley, and one of his contacts made a fortuitous introduction to Larry Page. After Lars tried for weeks to get a meeting with Larry Page, it was finally scheduled for June 4. Jens found out about the meeting only the week before. With holes in his socks and thirteen dollars in the company bank account, Jens booked a flight to California, using a credit card, to accompany Lars to the meeting with Larry Page and Megan Smith.

Early in the meeting Larry challenged the brothers. "Why don't you guys just do this in a browser?" he asked. Larry had a way of leaping over huge technical hurdles in search of the obviously superior user experience, and subtly belittling anyone that might suggest the obstacles in the way. You can do that when you, as a teenager, build fully functional printers out of Legos, as Page had done.

The Where2Tech's experience was certainly better than what

MapQuest or other web mapping services provided. Namely, it was fast. In fact, much faster than MapQuest. It was similar to Keyhole, in a way: prerendering the map tiles that a user might want before they request them. If you are looking at a map of 6102 Gaston in the Pemberton Heights neighborhood of Austin, Texas, it's quite likely that you might pull the map to the east or the west, so Where2Tech would go ahead and load 6100 Gaston and 6104 Gaston onto your computer. Like Keyhole, Where2Tech predicted what map data you would want next and loaded it into your computer's memory before you asked for it. But the preloading of map tiles and holding those tiles in memory and doing it in a secure way—that was something that could only be executed by an installed application with read-write access to a computer device's memory (and protections against data scraping). Keyhole had gained lots of experience dealing with hackers attempting to scrape and sell our data. For example, unusual spikes in traffic on weekends were often associated with attempts to steal our mapping content.

By 2004, however, Larry's question about "doing this in a browser" carried some merit. Internet browsers, pushed forward by the Mozilla Foundation with its browser Firefox, had evolved to handle more complex tasks in an effort to speed web-based user experiences. Utilizing a new set of techniques called Ajax (short for Asynchronous JavaScript and XML), Firefox allowed websites to call for data in the background while not disrupting what was on the user's screen. Offering faster user interactivity in a browser—while keeping the data secure—was no longer out of the question.

"Let us get back to you," Jens had replied to Larry. The brothers sequestered themselves at a friend's house in the Berkeley Hills to try and make their app function in a browser. With a clear challenge in front of them, they coded nonstop to try to get a proof of concept working. While Jens focused on design elements and making the palette more Google-friendly, Lars focused on speed. Stephen and Noel built an ActiveX control—a type of software code called a plug-in that allows you to run an application inside of the Internet Explorer browser.

Years earlier, Lars had faced a similar technical challenge while working for a company that managed a network of Wi-Fi hot spots. The company needed a way of remotely monitoring whether a hot spot was up or down. Each hot spot had its own web page, but finding out whether it was up or not had previously required that page to be reloaded to requery the status of the hot spots.

What Lars helped to figure out was a way of getting a tiny bit of data back—namely the status of the router—without requiring a page refresh. It was a capability allowed and enabled only by the latest Internet browsers, Internet Explorer 6.0 and Mozilla Firefox, using Ajax. This proved to be a major leap for Where2Tech—and ultimately for Google Maps. With this new process of loading data, map tiles could be fetched and cached in advance of the user requesting them. This preloading of mapping data would speed up the user experience, making it feel incredibly fast and responsive.

Three weeks later, a follow-up meeting at Google happened with Larry Page and Megan Smith. Where2Tech had been successful getting its prototype working as an ActiveX plug-in, giving it at least the appearance of a service running inside a browser. Larry was intrigued by the concept even though he knew an ActiveX-based solution wouldn't be a long-term answer. In June 2004, the Where2Tech team was offered employment contracts and came to work at Google.

The four employees set up shop in Building 41. With little guidance or direction, they continued refining their project with the goal of putting something out on Google Labs, the company's relatively hidden sandbox for new ideas and technical experiments. Stephen, Noel, and Lars headed back to Sydney (though they continued to work on the project), while Jens remained in Building 41 and was soon joined by Brett Taylor and Jim Norris.

They were given almost no direction from Google. To kill time, they installed a traffic light above their space in Building 41 and coded it to turn red when Noel was away from his desk in Sydney and green while he was there. Jens doodled sketches on his Wacom tablet, exploring ways that Google could represent dots on the map: cycling between a variety of stars, circles, and squares before settling on a pin icon that could mark a spot without obscuring what was underneath it. He also ditched the early designs of maps that were based on the colorful Google palette and settled instead on a calmer color palette for their new mapping project designs.

One of their first real tasks was to help complete a technical due diligence of another company Google was considering acquiring that summer: Keyhole. They met with Brian and Michael as a part of the vetting process.

Then, in October of 2004, the Where2Tech and the Keyhole teams were all thrown together. We were left independent of each other, and no one person was in charge, but the teams had agreed to integrate Keyhole imagery into Google's new maps as soon as possible.

On our second day at Google, John was scheduled to meet with senior engineering leaders, including Wayne Rosing. I was confident that we would get clear insight from the company's leadership about Google's plans for Keyhole. This meeting was going to be our true north, when all would be revealed with perfect clarity.

Rosing was a well-respected and seasoned software manager and engineer. He had served in engineering leadership roles across Silicon Valley, including at Apple and Sun, where he had been one of the creators of the programming language Java. At Apple, he was the engineering director on the Apple Lisa, the forerunner to the Macintosh. Now acting as Google's VP of engineering, he was widely admired internally as a thought leader in product strategy. All Google engineers reported up to Rosing—and Rosing reported directly to Larry Page.

It didn't seem unreasonable to think that the October 22 meeting with Rosing might answer the basic questions about Keyhole's role and function at Google.

Later that evening, I sat down with John in the bright yellow lounge chairs outside his office. "So how did it go? What's the plan? What did he say?" I asked.

"Well," John started slowly, "I'm not really sure."

"What do you mean?"

"Well, I'll tell you exactly what he said," John said, pausing for a moment. "I quote, 'We shouldn't fuck this up.'"

"What is that supposed to mean?"

"I walked him through the whole thing: our current product

offer, our product road map, who our customers are, our revenue, our sales projections. At the end of it, he asked me: 'So you guys have customers, right? And you have revenue, right?' I said, 'Yeah, we have customers and we have revenue.' Then he said, 'Well, I think the plan should be that we shouldn't fuck this up.'"

"We shouldn't fuck this up," I repeated.

"That was pretty much it," John said matter-of-factly.

It was at this point that I began to think that I was a part of some sort of Building 41 social psychology experiment, and that a team of psychology PhD students from Stanford was going to emerge from behind some mirrored glass wall to tell me about their research on cognitive dissonance and the effects of ambiguity on teams.

But as I reflect on it now, I am beginning to understand Rosing: His salty language made a lot of sense. Keyhole was a highly functioning team. We knew our stuff. We had a strong leader with years of experience in mapping. In fact, John had significantly more experience in mapping than anyone else at Google.

There was a different path for Keyhole that I think other functional leaders at Google had been considering before they met John: to break up Keyhole into Google's functional areas and make us a part of Marissa's search team.

It would have been an easy approach and is often the strategy when acquiring a company. The best way to ensure those from a newly acquired company get into the corporate boat and start rowing in the same direction is to break the team up into their functional areas, like engineering, sales, marketing, and operations.

In the case of Keyhole, I believe Rosing saw John's potential as a leader, not just of Keyhole, but of all of Google's new mapping efforts. Maybe instead the Googlers should get in the Keyhole boat.

I think "We shouldn't fuck this up" was Rosing's way of saying, "We have a strong team here with a lot of experience in an important new area for the company. Let's keep them together and build a new team under John and Brian."

John's title at Google would be general manager, Keyhole. For the foreseeable future, everyone on the Keyhole team would still report to John. John, like Marissa, would report directly to Jonathan Rosenberg. Bret Taylor and the mapping team would report to Marissa, who ran Google Search and Consumer Products (read: all of Google except the infrastructure). A distinct organizational border was drawn between the Marissa camp and the John Hanke camp. What could possibly go wrong?

On Thursday, October 21, four days into our first week, Google's first earnings report as a public company was scheduled. This would be the first time that the financial world would get a look inside the Google financials and the first chance after the IPO to see if this company really was making money.

Google answered with a resounding yes, breezing past Wall Street's expectations for the first of many occasions. The reason, of course, was that, even in 2004, the Google magic ATM, called AdWords, was spitting out revenue at a fantastic rate. Now it was public for all the world to see: Google's ability to generate revenue is unparalleled in the annals of business. It was the fastest company in history to go from zero to one billion dollars. Since going public three months earlier, the stock price had already raced from $85 to $140. After this first earnings report, Google stock jumped to $190.

The Keyhole team happened to be seated right next to the marketing and PR teams in Building 41. Wild cheers and applause resounded throughout the Googleplex. Even for those who had been there since the company's inception, it was their first look into Google's financials. I tried not to look at the after-hours stock price more than twenty or thirty times that day.

At the end of our first week at Google, the Keyhole team was invited to attend the weekly TGIF meeting for employees in Charlie's Café, a sprawling auditorium with an elevated stage at one end. (One of the most amazing facts I learned was that employee No. 38 was a full-time chef named Charlie Ayers.) A large Google neon sign glowed on one wall. With 2,500 employees, the company had quickly outgrown the capacity of Charlie's. Videoconference rooms were set up to accommodate the overflow.

For this gathering, the Keyhole team was front and center in the section reserved for Nooglers (*new* meets *Googlers*). Each chair was adorned with Noogler multicolored beanies, complete with red propellers on top. Snacks and beer and wine flowed liberally. Two chefs emerged from the kitchen carrying a surfboard over their heads; it doubled as a giant serving tray and was stacked with the finest sushi in Silicon Valley. Eric, Larry, and Sergey invited John, wearing his beanie, onstage to demo Keyhole EarthViewer for the company. As would any grown man wearing a propeller cap, John looked ridiculous. For the majority in attendance, it was the first time that they had experienced EarthViewer, and John's demo was met with audible gasps and thunderous applause as he piloted around the globe.

Eric Schmidt shook his head in disbelief as John left the stage.

"That is just incredible," Eric added. "And to think that he did that demo running off of a basic Sony Vaio laptop." As someone who had been in the computing industry for thirty years, Eric fully understood the technological step change that John had just demonstrated—and what it might mean now that it was a part of Google.

Immediately following the TGIF, the entire company filtered onto the lawn between Building 43 and Building 41, and posed for a special photo. An avid photographer, Michael Jones climbed up onto the roof of Building 43 and took a photo of the company from above that we quickly imported into Keyhole. By the time the Googlers returned to their desks, there was an email to the company from Jonathan Rosenberg officially introducing and welcoming Keyhole to Google and informing the Googlers that they now had access to Keyhole EarthViewer for free. They could simply log in with their @google.com email addresses.

The resulting surge of interest from Googlers almost crashed the Keyhole servers. It was a useful learning experience, if not a red flag. Making the service free—even for 2,500 people—would be met with exponentially greater usage and demand.

The following week, on October 27, five days *after* the Google earnings report, the announcement of the Keyhole acquisition was finally allowed to be made public. During the predawn hours, a small team from Keyhole and Google gathered at our old offices. Chikai brought Krispy Kreme donuts, and Dede had already brewed a fresh pot of coffee. Since Google's stock was now traded publicly and this was financially material news, the acquisition announcement was set to coincide with the market opening at nine o'clock Eastern time (or six o'clock Pacific time). It was a highly orchestrated sequence of tasks. As the marketing lead, I played the conductor.

Keyhole service server redirected to new Google servers at 5:45 A.M., check. New website pushed live by Patricia Wahl at 5:55 A.M. ("Keyhole's Feeling Lucky"), check. Google blog post pushed live 6:00 A.M. by Karen Wickre at Google, done. News embargo on all prebriefed press lifted at 6:00 A.M., yep. Press release launched on Businesswire at 6:00 A.M. by Noah Doyle, done. Email newsletter to professional customers 6:05 A.M. by Ritee, sent. Email newsletter to consumer customers 6:10 A.M., sent. Ed Ruben switched over the auth server to be a Google server, and monitored it closely due to anticipation spikes in downloads.

Finally! No more secrets from friends and family; we could talk freely about the acquisition. Emails from friends and family and business counterparts started coming in immediately with subject lines like "Kilday-WTF?" and "OMG. Call me!" and "Wow. Just Wow!"

My favorite phone call of the day might have been the one that John Hanke received from the landlord of 94 La Avenida. You remember, the one that I had listened to John haggle with on his phone four years earlier while we drank beers and ate tacos in San Diego.

As part of the acquisition, John and Dede had been required to track down and receive signatures of approval of the offer from investors that represented 60 percent of the outstanding shares of the company. For two weeks, they tried to track down our landlord to inform him of the acquisition and to get his election form completed and signed.

When John answered the call the day of the acquisition announcement, the landlord was frantic. "I haven't missed out on the deal, have I?" he asked. "What do I need to sign? When and where do I need to be?" Then the landlord explained why he was so hard to track down. He was calling from a hospital bed at Stanford University Medical Center. Ten days earlier, he had suffered a heart attack and was still recovering. That morning he was resting in bed, half paying attention to the morning market news on CNBC while he ate breakfast.

On the mounted television, the reporter said: "Search engine giant Google is announcing today their first acquisition since going public. Mountain View–based Keyhole, the tiny mapping software company behind EarthViewer, which you may have seen on CNN and other broadcasts. The same one that their landlord owns 13,000 shares in."

Okay, I did insert that last line. But truthfully, the news had almost given the landlord another heart attack. He spit out his coffee, jumped up out of his bed, and scrambled for his phone and John's number.

"You're fine. You're okay. No need to worry. Your stock is safe *and* is going to be converted into shares of Google. Get back in bed. We'll be in touch with you with more details later," John assured him, and the landlord calmed down.

But the landlord did have a final question for John. It was a common theme emerging from the multiple questions being asked by friends and family and press and analysts alike. "Google? Buying Keyhole? Can you explain what they plan to do with you guys? Because, honestly, I don't get it?"

"Well, we'll have to wait and see on that one," John said. "We'll just have to wait and see."

Chapter 10

A FRIENDLY WAGER IN BUILDING 41

One morning we arrived at Building 41 to find the cubes of Bret, Jens, Lars, and the rest of their team empty. At first we thought they might be running late, but then it became clear that the team was not coming to the office that day. Apparently, Bret and Marissa had organized a mapping strategy off-site—but had not invited John or Brian or anyone else from the Keyhole team.

For my part, I had been in several meetings with Bret Taylor where the force of his personality and intellect took over the meeting. With his deep baritone and cocksure ideas, he even trumped John Hanke for control of a meeting mindshare and of the Google mapping project. Since Bret held the most tenure at Google, his ideas were practical and grounded in a deep understanding of how to get things done at the company.

Google was full of Brets. Twentysomething Stanford graduates. Not just project managers and engineers, but also the guy who worked at the Techstop or the young woman in human resources who instructed me on how to fill out health insurance forms. Everyone was brilliant, ready to answer your questions in detail.

By the time I started in October 2004, Google was receiving

three thousand résumés a day (which works out to over a million a year). As a result, I soon found myself interviewing these bright, creative college graduates, most of whom never made it through the Google interview gauntlet. During one interview, I met a young man who presented nine marketing ideas for Keyhole EarthViewer—and his ideas were great. The candidate didn't make it to his second interview; another Googler dinged him for one reason or another (I never found out why).

Though Bret was new to the mapping industry, he seemed to me to prefer to figure it out on his own rather than tap into Keyhole's experiences. Bret and his team pursued contracts with data providers, sometimes with vendors that Keyhole already had relationships with. Daniel Lederman received calls from a handful of data providers asking about these new Google contacts calling to set up meetings. Engineers were being recruited to the separate teams, and Brian was competing with Lars about what projects a new engineer might be assigned to. We were quickly stepping on one another's toes. And egos.

Since I was on the Keyhole team, I technically reported to John but was often asked to help on projects for Marissa Mayer since she owned marketing at Google. Marissa had pulled most of Google's marketing functions into her power vortex, bypassing Doug Edwards and others on the marketing communications team for control of the marketing function (not including PR).

And Marissa started to become the face of Google at numerous public events and speaking engagements, including appearances on the *Today* show and *60 Minutes* with segments about Google, where she demoed Keyhole software as an example of innovation at Google, much to John's chagrin. He found out about the *60 Minutes* interview only because Debbie asked me to set up the EarthViewer demo station in Building 40, where the crew and Lesley Stahl would be interviewing Marissa.

For numerous reasons, I kept my distance from Marissa. First, of course, there was the matter of the little blue marble kerfuffle during the acquisition announcement. With a master's degree in computer science from Stanford, she was intimidatingly smart and intense, and surrounded herself with only the best and brightest from Stanford. Because I was a state-school grad and part of the Keyhole/John Hanke camp, I wasn't in her inner circle, a circle that she worked hard to create and curate in a way that was "her secret weapon," according to *Wired*. During her studies, she built a travel recommendation software, showing an early interest for searches with a geographical component. Marissa interned at SRI International and Ubilab, UBS's research lab in Zurich, Switzerland. After graduation, she received fourteen job offers, including one from Google in 1999.

In 2002, Marissa established a special college recruitment program to feed forty "gifted Googlers" into an elite training regimen called the associate product manager program, or APM for short. Once hired, these new college graduates were whisked around the world—often with Marissa at the helm—on an indoctrination into all things Google and marketing. I'd read wrap-up reports from these boondoggles, with fresh-faced Googlers standing on the Great Wall of China or under the Eiffel Tower just before their meeting at a Nike factory or the Louis Vuitton headquarters. Then these recruits were placed into six-month rotations in different areas of the company to cross-pollinate and learn various aspects of the Google business. This group of

Marissa loyalists was sprinkled throughout Google in product leadership roles. Not surprisingly, Bret and Debbie had both gone through the program.

I didn't hide my allegiances those days: John and I were inseparable as we enjoyed all the perks that Google had to offer. Free lunches at Charlie's Café. The Google basketball league. TGIF happy hours. The Google gym. If Marissa saw me on the Google campus—either in a meeting or elsewhere—she saw me with John Hanke.

To Marissa, a search that had a location component (for example, a hotel in San Francisco or open houses in San Mateo) was just another search. Whether it called for a result that was map-based or list-based, it was still a search. Given this, she claimed location searches as her territory and therefore the team that was building the best map-based results. Her desire to hold on to mapping, I can guess, was also driven by the fact that location-oriented searches were a part of a small percentage of searches that offered opportunities to generate revenue.

Consider this for a minute: Do you think searches like "average lifespan macaw parrot," "spell sycophant," or "David Hasselhoff drunk burrito video" are searches that advertisers are clamoring to show ads against? Uh, nope. As a member of the human race, you will be ashamed to learn that those types of searches are (I am not exaggerating) approximately 93 percent of all Google searches (you people!). Only 7 percent of all searches entered into Google return any ads at all (translate: Google only has the *chance* to make any money 7 percent of the time someone enters a search). In addition, Google chooses to not allow advertising for pornography.

As it happens, searches related to travel planning and real estate, both of which are ripe for map-based results, represent some of the most highly monetizable categories of searches done on Google. Marissa would have been keen to fence off this lucrative territory.

In December of 2004, only two months after the acquisition, matters became more complicated for me. Somehow I found myself responsible for the marketing of the soon-to-be-launched Google Maps as well as my original Keyhole EarthViewer responsibilities. Debbie Jaffe positioned the opportunity to me as a natural and somewhat low-involvement add-on to my responsibilities. Google Maps had not even launched. How hard could it be? I was glad that after only a couple of months of our working together, Debbie was asking me to take on additional duties.

After consulting John and receiving his approval, I agreed. Two people were listed on the worldwide Google Inc. org chart responsible for the marketing of Google Maps and the soon-to-be-released Google version of Keyhole EarthViewer: Ritee Rouf, Keyhole's marketing coordinator, and me. Suddenly I was in charge of marketing for Google Maps and Earth.

This meant that I was now reporting to both John and Marissa (through Debbie Jaffe). John had approved the move as he knew that Marissa controlled the powerful decision of which product was promoted on the Google home page. He guessed correctly that I might get Keyhole featured more often on Google's home page if I had a semblance of a working relationship with Marissa.

Every two weeks, a single one-line link shows on the Google home page below the single search box. Often the link points to philanthropic causes or is used to highlight a current event (for example, *Happy Bastille Day*). It may be only a single line of text, but it is likely the most valuable advertising on the planet, seen by hundreds of millions of people every day. And Marissa controlled what was shown there.

By joining Marissa's org, I was able, with Debbie's help, to get Keyhole slotted in for multiple rotations on the Google home page through the fall of 2004 and the spring of 2005; each time the promo generated tens of thousands of daily downloads of Keyhole EarthViewer.

Those double duties soon irked John, however, as I was now splitting time between two bosses. I was in countless meetings, day in and day out. Often I presented demos of EarthViewer to potential advertisers, like PepsiCo and Travelocity and Dell, as well as during many other meetings for Google execs, new recruits, and others. All of this meant that I was no longer always at my desk when John came looking for me.

Debbie ran marketing for all of Marissa's product lines: Google Local, a shopping service called Froogle, the photo-editing software Picasa, Gmail, Blogger, and likely some others I am forgetting. Oh, yeah, Google.com. Now that I was working with Marissa's team, I vowed to not be like the Blogger bros that included Biz Stone. The microsite publishing service team had been acquired a year earlier and seemed removed from the Google culture and refused to convert over to the brand for their product. While I became more integrated into Debbie Jaffe's team (and hence Marissa's org) on the marketing side that spring, Chikai quickly embedded into Bret Taylor's team (and hence Marissa's org) on the product side. After all, that team was in need of what Chikai specialized in—processing and serving a global mapping database.

Bret and his team's initial prototypes, based on Where2Tech's explorations, had used only a small sampling of some freely available government data sets. Now they sat among the cubicles in the middle of the Keyhole team, and our group had already assembled a library of mapping data over the past five years. We had also created the tools and process to manage the curating and importing of all of that data. Those late-night failed data import tasks completed by Wayne Thai, using the latest version of Mark Aubin and John Johnson's Earthfusion tools, had yielded a well-oiled data processing machine. (In fact, Mark delivered a tech talk to Googlers to explain the mechanics of the mosaicking tools being used in our software. Tech talks, a sort of brown-bag lecture, were almost a daily offering to Googlers, and just one element of the company's thoughtful efforts to create an environment that fostered communication between departments.)

A map is, I will remind you, essentially a database, where all the records have a location. Bret, Marissa, Jens, and Lars may have been developing the greatest music-playing application in history, but John and the Keyhole team had all the songs. Road network data, business listings data, border data, points of interest like parks and buildings, aerial imagery, and more. The Keyhole team was in a perfect position to jump-start the work that had been initiated by the Google mapping team.

Over a series of meetings in November and December, John, Marissa, and Bret began to realize this and settled on a common goal for the two teams: launch Google Maps in three months. The critical component of this success was hinged on one common denominator—*data*. And the one person with the keys to all the data was Chikai Ohazama. As much as John or Bret, Chikai led this effort and managed to find a way for the two groups to collaborate in harmony.

Starting in November of 2004, many people worked hard over this particular stretch of time, but I can honestly say that I have never seen anyone work as hard as Chikai Ohazama worked to integrate the Keyhole data into Google Maps. When I arrived each morning and I left each night, Chikai was there, coding away, meeting with server engineers and with Bret and the Rasmussen brothers. Saturday, Sunday, weeknights—he was *always* there. (Right before Keyhole was acquired, Google security established the policy that employees were not allowed to sleep at the offices. A recent college grad from Brown had been caught living at Google; he hadn't bothered to rent an apartment and had instead slept on an office couch, ate at Google cafeterias, used the campus Laundromat, and taken showers at the Google gym.) I often joked with Chikai that he wasn't allowed to live at Google. There was a pullout futon in his oversized cube, and he probably would have slept there if the policy hadn't been in place.

Chikai's diplomacy began with a friendly challenge to Bret and Jim and the Where2Tech team. The bet centered around a clear user experience that the team wanted to create. Future users of Google's new mapping service would have a fast experience, sped up by Lars and Bret's prerendered tile trickery. It also would have a button that would allow users to view the aerial or satellite image of that very location. The imagery database would be the exact same database behind EarthViewer.

With the help of a server engineer named Andrew Kirmse, Chikai set up a special cluster of servers and then enabled the Google Maps team to tap into the Keyhole imagery database. (John had actually worked with Andrew more than eight years prior on a gaming project called Meridian 59, widely considered to be the first 3-D role-playing community.)

Then Chikai bet the Where2Tech team that they couldn't integrate the Keyhole imagery database into the web-based Google Maps experience within a week. If they could do it, he would take them out to dinner at the restaurant of their choosing in Silicon Valley. If they failed, the team would owe Chikai dinner.

What Chikai didn't know was that Jens and Lars had been anxiously awaiting this day since they had predicted the integration with Keyhole aerial and satellite imagery after they completed the technical due diligence on our company. They, like Keyhole, had been waiting for the acquisition deal to finally close. Though Chikai didn't know it, the brothers had already built a proof of concept for how to integrate the data.

The Where2Tech team happily accepted Chikai's challenge, and they succeeded in less than twenty-four hours. By January 2005, the internal Google Maps project—which now had regular internal builds that were demonstrated and tested—included aerial imagery. As I looked over Jens's shoulder one afternoon, he took me through the latest version. I had never seen anything like it before. It was pure magic. It was a superfast map running inside a web browser that you could switch over at any point and easily view satellite and aerial imagery. It was so much faster than any other web-browser-based map out there. Only then did I start to gain an understanding of Larry's words: "You guys need to be thinking bigger." As I played with the alpha version of Google Maps late one night, I started to realize something: *This is going to totally leapfrog MapQuest.* It was clearly a superior user experience.

The early successes—and Chikai—brought the two teams together. He paid (with his own funds, not with his new G card) for dinner at La Bodeguita del Medio, an upscale Cuban restaurant in Palo Alto.

While adding aerial and satellite imagery to Google's maps presented a clear technical path, it did not provide a clear business path. We did not have the legal permission from our aerial and satellite imagery providers to use that imagery on a free web-based service. Companies like Airphoto USA and Digital Globe were rightfully concerned that malicious attacks from outside hackers could download entire libraries of aerial imagery. Daniel and John still had much work to do, along with Google lawyers, to renegotiate our data deals. It was a costly and time-consuming process. Legally, we could not permit the Keyhole imagery to launch with the first version of Google's web-based mapping products.

Somehow, though, Bret and Jens and team didn't seem to appreciate this enormous undertaking in front of John and Daniel. There still existed an us-versus-them mentality between the two groups. The Google mapping team seemed to think that Keyhole was withholding our highest resolution imagery for the EarthViewer product.

"You guys had your earn-out milestones that you were trying

to meet," Jens said to me a number of years later, "so we knew that you didn't want us to have the best imagery. You wanted to hold that back for EarthViewer." This is completely untrue, but good luck convincing the original Google Maps team of this.

Trust issues aside, there was a third feature of this killer new map being created in Building 41—an integrated Google search box. After all, while there is a wow factor to a fast, fluid map that includes aerial/satellite, it is only useful if you can search that map to find what you were exactly looking for.

Before Where2Tech or Keyhole ever set foot on the Google campus, there were several talented Googlers thinking about how to create the best experience for the users searching for a result connected to a location. This team was led by a tall, lanky software engineer named Dan Egnor.

Dan had come to Google by way of a contest. In April of 2002, Google held a programming challenge marketed as the First Annual Programming Contest. While the prize included $10,000 and a VIP tour of Google's Mountain View facilities, it was also a genius way of sparking new ideas. Contest participants were given access to data—900,000 pages of websites, to be exact—and told to frame their projects on exploring creative uses of this data.

On May 31, 2002, Google announced that Dan Egnor had won the contest with his project, which was simply titled "Geographic Search." Dan wrote a software algorithm that crawled the 900,000 web pages looking for street addresses. Then he took those addresses and geocoded them, which is tech talk for translating an address into a latitude and a longitude. With this specific kind of information, you can use that database to create a map. This meant that if a web page included a street address, it could be plotted on a map.

If Larry Page's mission for Google had been "to organize the world's information," Dan Egnor's new algorithm aimed "to organize the world's information *geographically*." He was offered a job in 2002, but didn't immediately accept since he didn't want to leave New York City. One year later, Google successfully hired Egnor as its first engineer in New York, and he set up an East Coast outpost for Google, working on a project with another engineer, Elizabeth Harmon, called Google Search by Location.

Harmon and Egnor productized the crawling of web pages looking for addresses and geocoding those web pages. Then they combined that data with other databases licensed from third-party data providers, such as InfoUSA and Dun & Bradstreet. In 2004, Google patented Egnor's work with the title "Indexing documents according to geographical relevance."

Prior to 2004 and Egnor's work, building and maintaining an accurate business-listing database was a Sisyphean task: You were never done with it. Traditionally, all mapping companies—from MapQuest and Navteq to TomTom and Keyhole—relied wholly on these data providers that created, updated, and delivered business address data. No one tried to create their own. Amassing the data required hundreds, if not thousands, of telemarketers calling businesses to verify the accuracy of the information. Even so, the data was notoriously inaccurate. There were thirty million business locations in the United States alone; existing businesses were constantly moving or failing, new businesses were starting, and new franchises were opening their doors.

Delays in importing the data added to the lack of accuracy. For example, Keyhole received its updates from InfoUSA every six months (on dozens of CD-ROMs). Depending on Chikai and Wayne's workload, we might get the database updated every eight or nine months (assuming InfoUSA telemarketers had correctly identified a change of address). MapQuest and others, including Keyhole, were notorious for directing users to businesses that had long since moved or had closed altogether. I remember late one evening arriving at a FedEx location, hoping to ship an urgent package, only to find a darkened store and a sign that read WE HAVE MOVED. After all, a map is only as good as the data it geographically represents.

Egnor and Harmon were now working to create the cleanest, most complete, most up-to-date geographic data for Google. If successful, we wouldn't have to rely solely on third-party data providers. Instead, we would use that data as a start, and then compare that data to the database created by Egnor and Harmon's indexing of web pages.

Think about the retailer Target. In 2017, the company had 1,792 store locations throughout the United States. This figure increased by about 300 more stores since 2007. Do the simple math: This means Target opened about 30 stores each year, and many others likely changed locations. One of the very first things that Target—or any other business for that matter—does when opening a new store or shutting down an old one is update its website with the new address. If you want your customers to find you, you keep your website current.

With Harmon and Egnor's work, Google didn't have to wait for InfoUSA to call Target to verify its address, update its business listing database, then ship that updated database out to customers, where this update waited in line with other database updates. Instead, the new database generated from Egnor and Harmon's website crawl would be compared with old data, and any old data that could not be verified would be marked with a warning of "this location may be closed."

In December 2004, a brand-new Target opened up in Sunnyvale, not far from my sister-in-law's house. When I searched "Target near Sunnyvale, CA" in an alpha build of Google Maps, the new store was the first result in the map, tagged with Jens's iconic pin on the new location in downtown Sunnyvale. For comparison, I did that exact same search on MapQuest and Yahoo! Maps. These two searches completely missed the existence of the new Target store and instead pegged a location on El Camino Real Drive, the site of the old store that had closed one year earlier.

"Oh, wow, you gotta check this out," I said, yelling at John from my office. He came next door to my office and I showed him the Target search, first on MapQuest, and then in the prelaunch version of Google Maps.

"I know, I just met with Egnor and Harmon last week," John said. "They came in from New York, and gave Brian, Daniel, and me an update on their work. It's going to be a major improvement over the old data supplied by the third parties."

Brett, Lars, Chikai, and team were creating the fastest, most beautiful base map, but what was most important was the data. The places of interest, or POIs, being plotted on the map. Egnor and Harmon's work was going to ensure that Google's POIs—our geographic information—were the most comprehensive and most current.

It was an integral element of Google's mapping initiative, which now included the Keyhole team. The mission of the company was to organize the world's information, including the information that existed in the real world with its physical locations. Egnor and Harmon's work extended Google Search from indexing web pages to indexing the world around us.

I should also point out the fact that for the Target/Sunnyvale search, I could type the information into a single box in the alpha version. I wasn't required to search in multiple boxes. On MapQuest, I clicked on a radio button called "businesses" and then typed "Target" in the business name box, and then tabbed over and inputted "Sunnyvale" in the city box, and then tabbed over again and typed "California" in the state box. Similarly, looking up an address required a user to click the address button, then "8006 W. 31st" into the street address box, then "Austin" in the city box, then "Texas" in the state box. As with all Google searches in Google Maps, there would simply be one box. This was such a foreign concept. (Hence the example query "Ice Cream in Poughkeepsie" that lived under the Google Maps search box for years.)

Egnor and Harmon's work surfaced first in a sort of precursor to Google's mapping products on a service called Google Search by Location. This service had existed primarily as a proof of concept as well as evidence that there was a true market need for a mapping solution. You could find the Google Search by Location service buried on Google's company website in a section of experimental projects called Google Labs.

By April of 2004, even before the Keyhole acquisition, Google Search by Location had been recast by Marissa as Google Local. Bret Taylor, Jim Norris, and another product manager named Thai Tran were on the Google Local team prior to Where2Tech and Keyhole arriving at Google. Though still in beta, Google began marketing the service to consumers and advertisers, moving out from the Google Labs obscurity to a link on the Google home page. As part of this rollout, businesses that were signing up to advertise on Google, using Google's AdWords service, could now opt to target their ads geographically, too.

By the end of 2004, all three teams were working together with a common goal of creating a killer map. We all brought something revolutionary to the table: Bret and Lars and team created the beautiful map view in a browser. John and Chikai and the Keyhole team brought the stitched aerial and satellite imagery view, now also in a browser. Dan Egnor and Elizabeth Harmon developed the best point data, representing the most current and complete database.

The three teams were being referred to as the combined Maps/Local/Keyhole team. Half the group was working on the launch of Google Maps and half on the Google version of Keyhole EarthViewer, which had yet to be named.

By now, the teams were beginning to blend more effectively. I particularly enjoyed the reserved Jens Rasmussen. It was a tremendous asset to have such a strong designer on the team. He drove a new red Ferrari, and on Fridays he brought in Danish pastries for the whole team. Lars was extraordinarily sharp, well respected, and insightful. The work the brothers did—with Bret Taylor and Jim Norris's help—was nothing short of amazing.

Admittedly, I did begin to wonder about whether this mapping service would start to impede the demand for our flagship Earth-Viewer software. It was that good.

John had a more nuanced view: He was excited about the Ajax technology too, but also knew that the website was light-years from being able to render the Earth the way that the native 3-D client application did. EarthViewer now included terrain for the entire world, 3-D buildings, and hundreds of data overlays. The Ajax-based website could show only static 2-D map tiles. (The technologies would eventually converge, but it would take nearly a decade to get there.)

Our last day of real work in 2004, Friday, December 17, Larry and Sergey unexpectedly handed out envelopes with $1,000 cash in them (ten crisp hundred-dollar bills) at the company TGIF. To the entire 2,500-person company. That night the annual holiday party was thrown. Google rented the entire Computer History Museum in Mountain View for the event. The museum was decorated in the theme of a deserted island, complete with a crashed Cessna 172 spilling out gold bullion adorning the entrance. At one point in the evening, a snaking conga line moved through the atrium to the steel drums of a lively Caribbean band. Hula dancers shook their grass-skirted hips. After all, the company had doubled in size and the stock price had already zoomed from $85 to $192 in just six months, so the spirits flowed freely.

Late in the evening, an early Google executive, who knew of my dual Maps-and-Keyhole responsibilities, pulled me aside. We stood on a balcony, overlooking the lively gathering. "You guys need to be very careful," he said. "Don't let her appearance fool you. She's a black hole that sucks up all energy and responsibility around her. And if Hanke isn't careful, if he doesn't watch his back, she'll suck up all of you guys, too." His wife, standing next to him, nodded knowingly.

He didn't say her name, but I knew who he was talking about.

"I know that she didn't agree with the decision to not have John report to her," I said, "but I think he's okay, and I think the Keyhole team is okay, because Wayne Rosing really sees the value we bring to the table. I think he's got John's back."

"Well, I wouldn't be so sure of that," he said, taking a sip of his cocktail and looking around. "I heard that Rosing is retiring."

Chapter 11

LAUNCHING GOOGLE MAPS MANIA

At 5:30 A.M. on a chilly morning in late January of 2005, the forty-person Keyhole team (we had already begun to multiply our ranks) quietly shuffled onto a sleek black shuttle bus outside of Building 41. Our Wi-Fi-enabled charter bus was one of the forty-five queued up around the Googleplex that morning, ready to transport the entire company five hours from Mountain View to the Squaw Valley Ski Resort near Lake Tahoe. This trip was an annual affair, starting in 2000. Google had rented out the entire mountain for two days for its exclusive use. (This year turned out to be the last before the company divided the trip into more manageable waves.)

When we arrived on the mountain, it was a glorious blue-sky day. After checking in to our luxurious rooms, most of the team hit the slopes. John and Brian had purchased several dozen neon-green GPS location logging devices made by Garmin. At the base of the mountain, near the ski lifts, we set up a table where Googlers heading up the slopes could check out a GPS device in order to log their ski runs. When they returned, we downloaded their data and sent them an EarthViewer KML file that allowed them to virtually relive their experiences on the mountain that day (complete with tracks and speeds recorded). Many of the GPS logging devices of the day had begun adopting John Rohlf's KML standard, making Keyhole EarthViewer the preferred tool for visualizing the data collected. It was likely the first time that consumers (and Googlers) could see their locations represented as dots on top of realistic maps of aerial and satellite imagery.

John hoped to use this GPS stunt to give other Google engineers a glimpse of the innovative work being done on his new team, and to potentially lure others to come and join the growing Google mapping efforts.

For many Googlers, the demonstration of GPS-tracked ski runs by John and the Keyhole team was their first introduction to Google's mapping initiatives. Several Google employees on the ski trip openly questioned our strategy and plans, with little understanding of what the teams were working on in Building 41.

One might think, in the weeks prior to the launch of Google Maps, that the entire company might have been anxiously preparing for the announcement. In reality, few in the company had any idea what we were up to or gave it much thought. To other employees, we were just another team going about our business, preparing for the launch of another product. There were no expectations outside of Building 41, and in fact I don't think anyone *inside* Building 41 had any idea of the mapping mania that we were about to unleash.

The launch date was slated for mid-February of 2005. The product magically combined three features: a fast, fluid, browser-based map; a huge database of aerial and satellite imagery; and a comprehensive Google search with up-to-date geographic data. Of these features, launching with aerial and satellite imagery proved to be the hardest component to deliver as it depended on rewriting partnership deals with companies like J. R. Robertson's Airphoto USA and Digital Globe. (As we all know, if something involves lawyers, it's going to take lots of time.) John and Daniel needed to renegotiate all of the contracts with our data providers; we would have to pay J. R. Robertson and Digital Globe significantly more money.

With a clear understanding that we needed to create a mapping service for the whole world, not just the United States, Daniel leaned heavily on Digital Globe. Daniel went to them with a massive ask. Google wanted up-to-date satellite imagery for the two hundred most populated cities on the planet.

Michael McCarthy, the head of Digital Globe commercial sales, worked for weeks with Daniel to scope out the enormous data licensing deal and give Google a price. After multiple rounds of negotiations, Daniel and John were satisfied with the terms. The negotiated price for Digital Globe satellite imagery was brought down to a reasonable price of one dollar per square kilometer of imagery, but it was a massive amount of data. The total price tag: $3 million.

John stepped into our office and shared his consternation and hesitation about the price tag. That said, he had been getting signals, including in our welcome meeting with Larry and his challenge to "think bigger," so he and Daniel decided to go for it. John made plans to take the Digital Globe proposal up the ladder for budget approval. It was the first time he would be going to Larry and Sergey to ask for money.

In those days, there was not a formal process at Google for reviewing deals. If you wanted money for a partnership, an acquisition, or a colossal amount of mapping data, you scheduled a time to go over to Building 43 to meet with Larry and Sergey. In fact, you likely scheduled the meeting with Larry and Sergey personally. Both founders went through ridiculous phases of not having administrative assistants. Apparently their rationale was this: They were busier than they wanted to be, and if they simply stopped using administrative assistants, then they wouldn't have so many meetings to attend. Voilà!

With some relief, after a few missed meetings, Daniel and John finally locked in a time with Larry and Sergey. They walked into Larry and Sergey's shared dark, cluttered office located on a sort of third-floor mezzanine in Building 43, open to floors below. Daniel and John stepped over parts of disassembled gadgets on the floor and dodged the robotic laptop on wheels that was running around the office, being controlled by Larry at his computer.

Larry continued to work at his desk as he looked up briefly at Daniel, who opened up his laptop for the presentation. Even before Daniel had fired up the PowerPoint, Larry and Sergey quizzed John and Daniel about the technical specifications of the Digital Globe QuickBird satellite: the speed at which it traveled, the altitude, the size of the sensor, the resolution, the size of each image, the number of daylight hours it could operate, onboard storage, when it launched, and how much it had cost.

Larry and Sergey debated about the total size of the Digital Globe database, accounting for cloud cover, flight paths, date of launching into space, and the like. Daniel only made it to slide two of his five-slide presentation before Sergey interrupted. John sat next to Daniel on the couch, silently taking in what was unfolding before him.

"Why so little?" asked Sergey.

"Why so little?" Daniel was confused. The $3 million purchase represented a tripling of the database of mapping aerial and satellite data. Keyhole had built that database out over five years. What Daniel was proposing was three times what we had taken five years to build.

"Is this all their content?" Sergey asked, even though he had done his own math and knew it wasn't.

"Well, this is all the content we need," Daniel responded.

"How big is their entire library?" Larry asked.

"For the whole *planet*?" John asked.

"Well, their entire database includes imagery of remote islands in the middle of the ocean, sparsely populated regions of Africa and Australia and Antarctica, the Sahara Desert, and a lot of other stuff that we don't need. It's something like eighty million square kilometers," Daniel explained. "We are asking for your approval to buy three million square kilometers, which would cost three million dollars. This is the imagery that we need."

"Why don't you acquire the entire library?" Larry asked.

Daniel and John looked at each other in stunned silence. Only the U.S. military had acquired the entire Digital Globe database. The robot continued to roam across the carpeted floor.

"Yeah, why don't you go back and figure out how to acquire the entire database," Sergey agreed. "All eighty million." The cofounders appeared to be egging each other on, ratcheting up the stakes in order to take things as far out of the box as possible.

That afternoon I walked with John and Daniel to Charlie's Café for lunch. Neither of them spoke.

"Wow, did it go that badly?" I finally asked.

John smiled at me, glanced at Daniel with a slight cock of his brow, and said, "I think we may need to change the way we think."

After lunch, Daniel called Michael McCarthy to deliver the news. "Your proposal was rejected," Daniel said to McCarthy, who was playing golf on a course in Colorado. "How quickly can you put together a proposal for the entire Digital Globe database?"

I couldn't believe it. I still didn't fully understand the scope of what Larry and Sergey had planned for Keyhole and Google's mapping initiatives. It didn't appear to be grounded in any economic reality that I was capable of grasping. *Where, exactly, are they planning to take this whole thing?* I thought. *Google's mapping products? Have they lost their collective mind? Are we supposed to lose ours?*

That next Monday, John informed the Keyhole management team about Larry and Sergey's response to the Digital Global contract. His overall message to all of us: Brace yourselves. Get ready. A tidal wave is coming. This certainly presented a challenge for Brian and Chikai about what would be needed in terms of servers and manpower, and even practical concerns about where all of this data would be stored. "We're going to need a bigger closet for all of those Digital Globe hard drives," Chikai said.

With the deal for satellite imagery expanding exponentially and the timing of the negotiations slipping out, there was no way that the aerial and satellite imagery could be ready for the launch of Google Maps in early February. As a result, Marissa and Bret made the decision to proceed with the February launch date—*without* the Keyhole aerial and satellite imagery database. We were disappointed that the Keyhole imagery was not going to be included as a part of this initial launch, but understood the decision.

"Just ship it" was a common mantra at Google in those days. There was a strong desire to be nimble and fast, and to not let perfection stand in the way of progress. Bret embodied this determined spirit and took the lead with the launch of this new product.

The launch date and time were set for Tuesday morning, February 8, 2005, at 9:00 A.M. (Pacific time). At this point, Google did very little in terms of marketing in support of new products, and even by those low standards, this launch was going to be accompanied by zero marketing fanfare. On paper, I had technically taken over as the marketing guy on the team, but there was no plan and no time to do any marketing, like a press tour or launch trailer video. Plus, my focus was largely on the Google version of Keyhole EarthViewer: the downloadable application, not the Google Maps website.

A mere one-paragraph blog post announcing the new service, written by Bret, was queued up on Google servers, ready to be pushed live for the 9:00 A.M. launch on Tuesday morning. Because of this relatively early announcement (Google engineers are notoriously late arrivers to work) and because the blog post had links to the new maps.google.com domain, Jens and Lars and the rest of the team decided to go ahead and push the site live Monday night, and simply rely on "security by obscurity" before the official launch the next day. (Read: We were counting on no one finding the maps.google.com site for about eleven hours.)

On Monday evening, around six o'clock, the team assembled in their nook on the bottom floor of Building 41: Bret, Jim Norris, Jens, and Lars were there. Noel Gordon and Stephen Ma had both flown from Sydney to work on-site those last critical weeks leading up to launch. Even though we were not launching with aerial and satellite imagery, Chikai was there. Andrew Kirmse was also there, having become the conduit between Google Maps and the Google server infrastructure team.

At that time, Google owned about 400,000 servers worldwide, multiple times larger than any other company. Google was revolutionizing the way data centers could be used as a platform for delivering all kinds of new web services with efficiency, redundancy, and speed, and redundancy. Kirmse was like the traffic cop for Google Maps; he helped figure out how to spread the Google Maps load across the globe to deal effectively with the traffic. With more than ten major data centers, he stood at an intersection that was about to become a lot more congested.

That evening Bret and Andrew pushed the buttons to launch Google Maps live to the world (though it only included a map of the United States). It was Monday, February 7, at 6:50 P.M. in order to make the 7 P.M. server push. (A new version of Google is pushed live every hour.)

Though I wasn't a part of their core team, I was excited to see the project go live. On my way out that evening, I stopped by to congratulate the team and had a quick chat with Jens. Their low-slung cubes were papered with maps from various historical atlases. Old-school globes hung from the ceiling. Rolling whiteboards were covered with computer code and other mathematical equations. A noticeable energy pulsed in the air. John was there and took photos of Lars, Bret, Andrew, and Chikai. Together, after the buttons were pushed, we watched the initial usage charts start to show web traffic (as was expected). That night, as I left Building 41, Sergey walked past me with a bottle of champagne and some red Solo cups. Though it was a low-key launch, many Googlers had been testing the new Google Maps service for two months internally, including Sergey. He had become an early tester and champion of the service, and was a regular visitor to the team during those weeks leading up to the launch.

The team hung around the office together and ordered a few pizzas while watching the servers for any performance issues. At around 8:00 P.M., traffic started to pick up on the site. The URL had been discovered, and a few people were starting to play with Google's curious new innovation.

At about 7:45 A.M. the next morning, Google Maps was "slashdotted," meaning that someone had started a thread on the influential tech website Slashdot (slashdot.org) about the slick new mapping service buried deep in the Google Labs site. Traffic ticked up again, and by the time I arrived at the office at nine o'clock, even before Bret's Google blog post had gone live, Google Maps exceeded the traffic forecast, and server allotment, for the entire twenty-four-hour period. Kirmse and the Google server infrastructure team started getting nervous.

The initial feedback to Google Maps was one of surprised

delight. On forums like Reddit, excited users commented about the ability to grab and move the map, and not having to wait for the map to refresh. The experience was so much faster and more fluid than any other mapping service, and this speed led to more free-form exploration and discovery.

That exploration was, however, limited to the United States. As quickly as the positive feedback came in, international users began complaining about this limitation: You could zoom back out to a point of view where you could see only the United States surrounded by oceans. The team didn't even bother to render the outline of countries other than the United States. It was like one of those alternate reality maps on the cover of *The New Yorker*, and in it, Google was essentially saying that the rest of the world did not exist.

Internally at the company, the success of the launch of Google Maps set off both excitement and immediate pressure to make the service global. The most frequent comment was that Google once again had prioritized the United States over every other country. Bret, Jim, Jens, and Lars had built a brilliant front-end access point into a world of maps; now this tool needed to be extended internationally.

Soon country managers throughout Google began lobbying executives in Mountain View for the service to be rolled out in their respective country first. I was being worked over by marketing managers on Debbie's team that were sprinkled around the globe. Suddenly I was very popular across her Google marketing org. "Hola Guillermo, congratulations on the amazing launch!" wrote the regal Spanish country manager Bernardo Hernández in an email. "As the number three country for Google in Europe, with 62 percent market share, I sincerely believe España should be on your short list for consideration as John Hanke develops a master plan for a global rollout. Can you please help me make the case to Señor Hanke to consider Spain? PS: We would love to host you and John in Madrid if you ever want to visit Europe." Then he offered a list of contacts with data providers, press, and potential advertisers and efforts he was willing to make in Spain in order to support a rollout.

This topic came up more than once in TGIF meetings presided over by Larry and Sergey. On one particular Friday afternoon, Brian was pulled up onstage to field a question about when we were launching in Africa. He performed an often-repeated tap dance about how hard the team was working to expand Google Maps globally.

Keep in mind, rolling out Google Maps in a new country was more of a "business lift" than a "technical lift." It wasn't new technology in a new country. It was the exact same technology, but with new data. It translated into multiple deals with data providers and partnerships and even potential new companies to acquire. Budgets, deals, negotiations, schedules, prioritization. This was the kind of work that Keyhole had been doing for the past five years, but on a much smaller scale. John and Daniel and Chikai's data processing tools and team moved into the mapping driver's seat as the company aimed to quickly roll out Google Maps internationally.

Japan was the next country to have Google Maps go live, in July of 2005. An entire set of data provider deals was completed to get the service up and running, spearheaded by a determined product manager named Kei Kawai, from Tōhoku, Japan, who had contacted John and promised to work tirelessly through all of the partnerships and data deals needed to launch Google Maps in his home country. Google Maps in the UK was up next (led by a dynamic rising head of European marketing named Lorraine Twohill), and then Ireland and France. John prioritized a list of countries, and Daniel Lederman started hiring business development people around the world. In our shared office, he kept a list of countries and the data providers in each of those countries on a whiteboard. This list also started including entire companies that Google might buy to advance its efforts to map the planet.

The initial launch of Google Maps in the United States was a runaway success. But because we were required to redo all of our satellite imaging Keyhole contracts, we had not included satellite imagery at launch. Chikai and the Keyhole imagery team were disappointed to not have a part in that initial tidal wave of global interest. Had we missed the boat? Chikai worked even harder, desperate to launch the database of aerial and satellite imagery as a feature of Google Maps and join in on the success. With John, he targeted a release date of mid-April of 2005.

Daniel finalized the multimillion-dollar satellite imagery deal with Digital Globe to license all of their data, and soon a whole floor in Building 41 was filled with disk drives of satellite imagery for Wayne Thai to process. (The multiple terabytes of imagery were too much to transfer over the Internet, although there was a plan in motion to transition the delivery over an upgraded fiber-optic connection.) We did have to make a concession to appease the Digital Globe's concern about data scraping. In a web browser (meaning Google Maps), we couldn't enable users to zoom all the way in to the highest resolution of imagery. Google Maps would need to clamp the resolution one zoom level above the highest resolution detail. In addition, we needed to watermark the imagery faintly with a copyright notice of "Copyright Google 2005."

John and Daniel had little control over this artificial restriction of the zoom level, but once again the core Google Maps team perceived the situation differently. Jens, Lars, Bret, and the rest of the team continued to think that Daniel, John, Brian, and Chikai were holding back—that we were reserving our highest-resolution, best imagery for the upcoming release of the Google version of Keyhole EarthViewer.

As we accelerated toward the launch of aerial and satellite imagery on Google Maps, a key naming issue remained unresolved. *Aerial* had been dinged because much of the imagery was from satellites. *Satellite,* however, was also technically incorrect, because much of the imagery was taken from airplanes. *Satellite,* though factually inaccurate, did offer this advantage: From a marketing standpoint, it was the sexier choice. Users wanted to believe that they were looking at Earth from space, as if they were operating a surveillance satellite.

At the last minute, Bret surprisingly asked me to figure it out via email. He was just too busy, but I think he was also too tired to fight about it. "Can you take care of this and let me know what the consensus is?" he wrote.

I was happy to walk to the halls, gather some final input, lobby a bit, and get a final decision from John. (I did ask Marissa, but she deferred the decision to John, since it was the Keyhole database.) I was happy to deliver the final decision to Bret, as it was the name he too wanted, even if it was somewhat incorrect. It would be *Satellite*.

The launch of Satellite Imagery happened on the morning of April 4. Once again, despite reviewing launch traffic projections, we were overwhelmed by the popularity of the new service.

The appeal of the service was part voyeurism, part useful. You now had the ability to view your house, your neighbor's house, or your ex-girlfriend's house. It was too tempting to resist. On the day of the launch, usage of Google Maps started creeping up, and the team expected that at some point it would start to taper off. But the traffic curves didn't taper off. In fact, as the day turned to evening, the traffic only increased.

Chikai was still at work in his cube at nine o'clock in the evening when Sergey and a group of server engineers appeared. The Satellite Imagery launch was so popular and server traffic loads so high that the new product was threatening to affect the overall speed of the Google home page. The server traffic from the launch of Satellite Imagery on Google Maps *tripled* the traffic load from the initial launch. This shouldn't have been a surprise. Despite our efforts to expand Google Maps to additional countries, it still didn't cover the entire world. Digital map data simply didn't exist for large portions of the world. But the Keyhole-produced imagery data set was truly global. Thanks to the expanded partnership with Digital Globe, the satellite portion of the map extended everywhere—throughout Africa and Latin America, and to the most remote parts of Asia and Antarctica. It was the first global map to be offered online in a browser and the first time this kind of image-based view of the entire world had ever been available through a web browser free of charge.

The BBC wrote "Google Maps Gives Fresh Perspective," and *Wired* wrote "Google Maps Is Changing the Way We See the World." *MIT Technology Review* added, "Even on the surface, it's clear that Google Maps goes much further than older interactive map sites. The stunning satellite views, along with the ability to drag the map in any direction without having to wait for the page to refresh, are the most obvious advances."

For anyone who had ever chosen to sit in a window seat on a flight, there was now a website that mirrored this experience: Google Maps with Satellite Imagery. And it could be used in a browser, without having to sign up for an account with Keyhole EarthViewer (still being sold for $29.95).

For the Keyhole team, we were ecstatic. We had hoped that the launch of satellite imagery of our Keyhole database on Google Maps would be popular. As the new kids on the Google block, we truly wanted to contribute to the larger Google—and Google Maps—mission. But no one predicted what a hit it would be. While we had been only tangentially connected to the February launch of Google Maps, we represented the core of the April launch of Google Maps with satellite imagery.

Finally it was becoming clear to us why Keyhole had been acquired. Google was launching a moonshot mapping effort to transform how we found our way in the world—and we were going to be at the center of that moonshot.

This launch formula would be repeated around the globe: We would first launch Google Maps in a country, and then a few weeks later would add in satellite mode. The satellite mode always attracted millions of new users, creating a huge spike in demand. Inevitably, this jump in traffic from the satellite imagery tapered off, but the constant level of traffic settled at a much, much higher usage level.

The satellite imagery was a great marketing hook for Google Maps. This imagery generated interest and brought in new users, and those new users ended up sticking around because the Google Maps product was so superior to the competition.

A week after the launch, Marissa sent out a company-wide email. Google Maps with Satellite Imagery, she was happy to report, was the most successful product launch in Google's young history, far outpacing the previous biggest launch: Gmail. Marissa recognized Chikai for his role in the launch and all the weeks of his hard work leading up to it. Within weeks, he was summoned to her office for a handsome reward. No one knew what it was, but I sure did like the way the leather seats felt in his new black Maserati.

Marissa sent out another email a month later: This one announced a decision unpopular among many on the team. The name of this wildly successful new product was changing: Google Maps and Google Local were merging into one product. And that product would be called—drumroll, please—Google Local. Google Local?

"What the hell is a Google Local?" I yelled out to John when her email arrived in my inbox. I jumped up and ran into his office. John swung around his chair and said, "I know, that's a good question." He hated the name as much as I did. John reminded me that it wasn't his decision. Bret and the entire Google Maps team still reported to Marissa, and she clearly wanted to combine it with the product that she had originated: Google Local.

Rosing no longer provided air cover for John from Marissa.

Like many to follow, Rosing left Google in May of 2005 after the IPO with his millions in search of new engineering moonshot challenges, which for him meant building software in the Atacama Desert to detect and intercept asteroids with planet-defending lasers. I'm not making this up.

Marissa had her reasons for the name change. For one, she believed that the term Google Maps was somehow limiting; that it would be perceived by users as only about finding driving directions, not local businesses. "The product is so much more than just a map," Debbie argued during one of her weekly marketing team meetings.

Debbie and I had a frenemy relationship: She was every bit as good of a personal friend of Marissa Mayer's as I was of John Hanke's. She predictably channeled Marissa's position, and I predictably channeled John's.

I argued that it would be a lot easier to teach Google Maps users to search for businesses than to try to teach users to go to something called Google Local. My position was based on a rule from visual artist and author Wendy Richmond, a rule that I have always considered a foundational marketing truth. It goes like this: "If you want to teach someone something new, start with something they already know." So this was my line of thinking: People know what a map is. We should start with Google Maps and expand their understanding from there.

Through Debbie, Marissa argued back that there was another reason for her decision: In 2004, a "local ad" option in Google AdWords had launched, and in its first year of existence the option attracted considerable business. Advertisers could easily add the "local ad" option onto their Google ad campaign with a single checkbox: Your Google Local Ads would immediately start appearing on the product called Google Local.

A pattern emerged as Google Maps rolled out globally. We would launch as Google Maps in a country; that product would soon get local business data and advertising added to it, and then the product name would change from Google Maps to Google Local.

Working out together in the Google gym one late afternoon, John reminded me that though he didn't like the decision, it was Marissa's decision—and Marissa's alone—to make. Young Googlers filed into the free yoga class that was starting up in the nearby mirrored studio. Despite my concerns, John told me to back off. "You need to pick your battles," he said, talking me down from any sort of pursuit of this naming issue. This was not a fight John was willing to pick right now. Marissa kept her grips on the Google Maps—or Google Local, I should say—team. For the time being, John kept his distance.

Chapter 12

SPARKING A NEW INDUSTRY

That spring of 2005, I was recruited by the head of the Google Book Search team, Dan Clancy, for a very important role: center on his Google basketball team. The league played on a set of hoops installed on a black asphalt parking lot outside of Building 45, and Dan and I had played a lot of pickup basketball on that court after work.

From my days back in Boston, I knew there existed an inverse relationship between basketball skills and academic pedigree. As I headed out the door to go play on Harvard's courts those years, I would often quote Matt Damon's line from *Good Will Hunting* to Shelley, "I'm going over to Hahvahd to beat up on some smaht kids." This followed me to Google: I was one of the better basketball players in the company.

Clancy, who had joined Google in 2004 after heading a team of five hundred software engineers at NASA Ames Research, had a PhD in computer science from the University of Texas. At Google, he now headed the ridiculously ambitious project to scan, index, and make searchable all of the world's books. At NASA, his software controlled robots fifty-four million miles away. Now his software controlled robots that flipped pages, scanned, de-warped, and read the contents of entire libraries.

I convinced John to join the team (he played in high school and was quite good); we decided on the team name of the Longhorns. Dan also invited Jonathan Rosenberg and a lanky software engineer named Jeff Dean to play with us. Warming up for our first game, I asked Jeff how long he had been at Google. Rosenberg stopped bouncing a ball, and everyone stopped and stared at me. Rosenberg said, "You really don't know anything about Google, do you, Kilday? Jeff Dean is the reason we are all here." Jonathan, Dan, and several others on the court began listing the acronyms and projects for which Dean was credited: Bigtable, MapReduce, Spanner, distributed computing infrastructure tools that all Google services are built on top of. Larry and Sergey may have hand-fabricated a Model T; Jeff built the Ford Motor Company. It was Jeff's code that allowed Google—and services like Gmail and now Google Maps—to scale exponentially.

Our warm-up chatter was often filled with Google mapping status updates for Rosenberg. Before the launch of Google Maps, I remember Rosenberg calling out the future one day, as we waited for our game to start.

"I love the new maps," Rosenberg said. "I think it's going to be awesome. But what I *really* want is a map that shows me all of the real estate open houses in Silicon Valley this weekend. And then I want to send that map to some device that I can take with me in my car so I can navigate to them." Keep in mind, this was still two years before the launch of the iPhone.

These mapping strategy conversations on the court with John

were short-lived, however. During our third game, as I was inbounding the ball, John cut quickly to get open. There was an audible snap as his Achilles tendon tore, recoiling up into his calf like a rubber band. He hobbled to the Google nurse on campus in Building 41, and Noah Doyle drove him home in East Bay. John spent the next months lurching from meeting to meeting on crutches and then later a walking boot, and I often followed him through Charlie's Cafe, carrying both our lunch trays to our table. I also delivered a trophy to John that summer: "Google Basketball Champions 2005: Longhorns."

After the launch of Google Maps, Bret and Jim Norris turned their attention to a brand-new product development effort: which would end up being one of the most important reasons that Google Maps exploded in popularity. One could argue that it was the single best marketing program ever undertaken for Google Maps, and even though it was essentially what Rosenberg had asked for, I didn't see it coming.

In Google's short history, the company had worked hard to foster positive relations with the software development community. Since Larry and Sergey were developers themselves, they appreciated firsthand companies that encouraged this kind of innovation by opening up their data and tools, usually with a set of application program interfaces (APIs). A Google Search API had been published to allow developers to create customized search services on other websites.

When Google Maps launched on Tuesday, February 8, besides the glowing reviews from big-name publications and users alike, the product also became a huge hit with web developers and software engineers wanting to modify those maps. In other words, they wanted to create their *own* versions of Google Maps, using their *own* data. Just as Rosenberg predicted on the basketball court, the first, most popular example of this was a real-estate-related map created by an animator working at DreamWorks named Paul Rademacher.

In late 2004 and early 2005, Rademacher had been trying to find affordable rental housing in the San Francisco Bay Area, an impossibility by most estimations. He searched for months—pouring over postings on Craigslist, printing out the listings and the map of the specific rental units, and then setting out on weekends in search of an apartment. His search had been inefficient and unsuccessful. As he drove around San Francisco, with these dozens of different maps printed out weekend after weekend, he thought to himself: *It would be so much more efficient if there were just* one *map, a single map showing all the available apartments, so that I could narrow my search geographically.*

The launch of Google Maps, with its advanced use of JavaScript and XML, cleared a path for Rademacher. Within hours of going live, he dug into the Google Maps code, trying to reverse-engineer its Ajax. His goal was to overlay *his* data: geocoded lists of available housing that he scraped from the Craigslist website. By Thursday of that week, only three days after the Google Maps launch, Rademacher posted *his* version of Google Maps, on a URL that he registered: housingmaps.com.

By Thursday night, Rademacher's map was already spreading rapidly, with thousands of San Franciscans taking his new map mash-up, housingmaps.com, for a spin. On Friday, the site made the rounds inside of Google too, with more than one person on the Google Maps team piling onto an email thread with the idea that we offer Rademacher a job. In fact he was offered a job (though it took him almost a year to accept). Immediately, dozens of copycat housing maps sites popped up around the country.

A second email thread circulated the Google Maps team that Friday with a link to another map mash-up, this one of crime statistics in Chicago. A web developer and musician named Adrian Holovaty had mixed the publicly reported stats of crimes in Cook County with Google Maps to create Chicagocrime.org. And *that* site spread rapidly, too. Once again, dozens of copycat crime statistics sites sprung up around the country.

Bret and Jim immediately recognized the populist web mapping publishing movement had been sparked, but now it was spreading uncontrollably. They realized that this posed a serious risk to Google infrastructure and data. These map mash-ups had broken in the back door of Google Maps, and dozens of eager web developers with some GIS data were rushing in. Google had no control of who got in and had no ability to weed out the bad actors. Most important, these mash-ups could easily break with any new Google Maps updates.

Bret and Jim closed the back door and swiftly developed an official Google service that would empower web developers with a predictable, documented tool for the creation of more map mash-ups. Access to the Google Maps API would be controlled: Developers could sign up for a Google Maps API key, which was similar to a token or a ticket that allowed you in the front door in an organized manner so you could use Google Maps as the base map of your mash-up. The official Google Maps API launched in June of 2005, allowing Google Maps to be integrated into anyone's website.

Did I mention that it was free?

Now, I must admit, this whole free thing was something I just didn't get. It was one idea to create a web service like Google Maps and launch it for free. This strategy was a part of a tried-and-true Silicon Valley business plan: Create a product, attract an audience, and sell advertising to companies wanting to reach that audience. It wasn't rocket science.

But what Bret and Jim created allowed *other* people to create *their own* products using *our* maps. Then those maps were featured on their own websites where they could sell advertising and generate revenue. To me, there appeared to be widespread skepticism about *any* moneymaking ideas for Google Maps and other Google services, especially among the engineering ranks.

Not only that, Google didn't get to aggregate and use the data that was being unlocked with the Google Maps API. I complained to Bret one afternoon about this, leaning over the partition of his cubicle. "Can we at least put something in the terms of service of the Google Maps API that would allow us to use their data on our maps?" I asked. When registering for a Google Maps API key, all developers were required to agree to the Google Maps API terms of service (TOS). I could already see the creativity of the community and the innovative ways our maps were being used as a base layer. It would be reasonable for us to be able to use that data on *our* maps, too. In that way, you could come to Google Maps and do a search for apartments in San Francisco or crime in Chicago or elementary schools in Austin or restaurant reviews in Portland. Google Maps could have become one centralized map hub (even for obscure data).

Larry and Sergey had similar thoughts, and poked around with Bret and Marissa about the prospects of claiming some sort of right to use the data that was going to be overlaid on top of our maps. But at Google, at the lower levels, there seemed to be a somewhat libertarian approach toward all data—and the idea was met with resistance. The general feeling was that our information, including our mapping databases, should be released from back-office silos and published by burgeoning masses of web developers with access to this geographic data. Bret embraced this libertarian spirit, and the API was launched without the ability for Google to use that data. Standing in the way of this freeing of data was perceived by many as standing in the way of Google's mission to "organize the world's information and make it universally accessible and useful."

Countless Google Maps mash-ups did end up shining a geographic light on all sorts of data that had never before been visualized on an easy-to-use, interactive, fast and fluid map. Examples abound: maps showing new "clean coal" (ha!) proposed projects in Texas, police brutality maps in Los Angeles, logging operations maps in the Santa Cruz Mountains, strip-mining maps in West Virginia, bicycle accident maps in Portland, Oregon. Bret and Jim set in motion an explosion of mapmaking creativity and a democratization of GIS data.

So many new Google Maps mash-ups were surfacing every day that I soon gave up trying to keep track of them all. There were even websites dedicated to tracking this rapid digital proliferation, like Mike Pegg's Google Maps Mania site that documented this dam break, posting four or five of the top mash-ups from the hundreds being created daily.

In addition to the thousands of personal API mash-ups, another class of Google Maps API developers soon materialized. And these guys meant business. These weren't the young, independent web developers looking to illuminate a particular issue, social cause, or database. No, I'm talking about real businesses. Making real money. Full-on commercial enterprises were born, companies that depended on *our* free Google Maps API. We did all of the heavy lifting to create the world's best base map and then gave away all that work for free. It spawned dozens of full-fledged multimillion-dollar (even multibillion-dollar) businesses. Businesses that you likely know, use, and love.

Yelp. Zillow. Trulia. Hotels.com. Strava. And later, Uber and Lyft and more. A parade of new Web 2.0 location-based services launched in 2005 and 2006, all using Google Maps as their foundation layer. Suddenly maps became a popular area of innovation, and a new type of venture-capital-funded start-ups, called location-based services, took off. And Bret and Jim's Google Maps API made all of these start-ups economically feasible. Google Maps API was a giant tech incubator, and we didn't ask for equity. Heck, we didn't even ask them to pay rent. It just wasn't in Google's DNA.

Eventually, several years later, the Google Maps API would be something that Google would charge for. It was actually something that the location-based service industry asked for, if not demanded. While they loved the idea of getting the ultimate base map for free, they didn't like the notion that Google could choose to change the service at any time or start showing advertisements on their Google Maps mash-up. For example, American Airlines featured a Google Maps–powered flight tracker on its website, and there was nothing that would have precluded Google from selling ads for United Airlines on that Google Maps API–powered mash-up.

Between the success of Google Maps, the Google Maps API, and the dozens of location-based services it enabled, mapping suddenly erupted as a white-hot tech trend in 2005. Success attracts competition, though, and in late March, a *Wall Street Journal* story circulated through the digital corridors of the Googleplex: Microsoft was jumping into the mapping arena.

The March 28 article was entitled "In Secret Hideaway, Bill Gates Ponders Microsoft's Future." Reporter Robert Guth was granted unfettered access into what Gates referred to as his "Think Week," his escape into the woods of Washington State to ponder the future of technology trends and Microsoft's product road maps.

According to the article, of the reported three hundred papers Gates might have read during this seven-day retreat, he chose to highlight one idea for the reporter. A product proposal titled "Virtual Earth."

Standing at his desk with ink-stained hands, Mr. Gates flipped through a 62-page paper titled "Virtual Earth," covered with his notes. It described future mapping services that deliver travel directions with live images and details on traffic conditions and other information. . . . "I love the vision here!"

And just in case anyone at Google didn't fully understand the point, the article continued:

In the MapPoint unit, source of the "Virtual Earth" paper, Mr. Lawler, the general manager, called a meeting to brainstorm on Mr. Gates's comments. . . . But word of the endorsement of the paper's overall vision had spread across Microsoft, and several other groups including Microsoft's research arm are now involved in the project.

Well, okay then. Thanks for the heads-up, Bill.

The story troubled John because, prior to the Google deal, Keyhole had had a brief relationship with Microsoft. The head of Microsoft's developer relations team, an executive named Vic Gundotra who would later play a prominent role at Google, had discovered Keyhole. He had shown EarthViewer around Microsoft and even touted it at a Microsoft developer event. Keyhole engineers had been dispatched to Redmond to optimize the code for the Windows operating system and Keyhole had presented its vision to Microsoft for an "earth browser" overlaid with data.

Two weeks later, Gates demoed Microsoft's Virtual Earth himself at the All Things Digital conference organized by Walt Mossberg of *The Wall Street Journal*. After the demo, Mossberg opened the discussion up for questions. The first three questions asked were not about Microsoft's Virtual Earth, but about Google Maps. By the third question, Gates was irritated. "Yes, Google is still perfect, the bubble is still floating, and they can do everything. You should buy their stock at any price," he answered sarcastically.

The Gates article and demo ended up having a consolidating effect on the somewhat disjointed mapping efforts inside Building 41. Emails with words like *accelerated* and *urgent* and even *unified effort* began making the rounds. Budgets—already swollen from early Google Maps successes—were doubled. And then doubled again. Head counts and new revised head counts were approved. (The Keyhole team started at twenty-nine; six months later, we were at two hundred.)

Many executives at Google had been outflanked by Gates and Microsoft in previous roles at companies like Netscape and Novell. In 2005, Gates had for some reason shown his hand. Maps were suddenly a battleground state. The threat of a 3-D client application from Microsoft to compete head-to-head with our upcoming launch of the rebranded Keyhole EarthViewer loomed large. Development was accelerated and features cut. John and Brian moved the launch date up a month.

Those first few months at Google, I think John was finding his footing within the company hierarchy. He was the leader of the Keyhole team and stayed in his organizational swim lane. But the Gates article prompted a scare inside of Google. This was likely the moment, I believe, that John seized the reins on Google's mapping efforts. As a response to the *Wall Street Journal* article, John and many of us on the team crafted and sent a detailed email to Larry, Sergey, and Eric, laying out a strategy and a tactical game plan for all of Google's mapping services. It included an analysis of the threat and spelled out likely Microsoft moves and supplier partners. Based on Gates's demo, John even proffered a guess of the aerial and satellite imagery data provider relationships that Microsoft had secured, and a description of those aerial and satellite imagery companies' capabilities and pricing. John set forth a plan for Google to speed the launch of the rebranded Keyhole EarthViewer. He called for unifying the Google Local/Maps/Keyhole teams. John recommended that Google invest heavily in data acquisition and server infrastructure backbone for our mapping services. In addition, he listed potential acquisition targets to pursue: regional mapping companies in important markets that Google should acquire to bolster our data acquisition and tech talent efforts.

By early summer, an agreement emerged among the Google executive team: It was going to take a massive effort to win in mapping. Larry announced his decision via email to all@google.com: He was establishing an entire new product line called Google Geo. All of Google's mapping efforts would be a part of this new Google Geo team. The Maps/Local/Keyhole separation would no longer exist. This new group would be headed by one leader: John Hanke, product director for Google's newly formed Geo division. All Google Geo engineers would report to Brian McClendon.

I found out about the move like everyone else—Larry's email. It was just like John not to say anything about this promotion. John told me that it was Megan Smith, our champion in corporate development, who had been lobbying for John with Larry and Sergey.

"They could have gone with Bret," he told me years later. "I mean, I'm sure he could have done it. He was certainly capable. But he *was* pretty junior at the time, so, I don't know. Whatever the reason, Larry and the rest of the executive team decided to give the job to me."

Bret, and the Google Maps and Local teams, would now all report to John. John would report directly to Jonathan Rosenberg. Marissa was officially out of Maps (for now). Bret, Jens, Lars, and many others didn't agree with the decision. From their perspective, they had created Google Maps—and deserved to own its implementation across the globe.

Lars stayed on the Google Maps team but relocated to Australia, founding Google's engineering presence there and laying the groundwork for a new project. Jens continued on in Google Maps in a design and front-end engineering capacity. Noel Gordon and Stephen Ma stayed on the maps projects, too: Noel helped me on a marketing project to put Google Maps on the seat back monitors of all JetBlue airplanes. And then Virgin Airlines. And then Frontier. And then several others.

Noel and Stephen were the first engineers in Lars's new Google Sydney office, and within a year moved off maps altogether. The Where2Tech team had only come together less than two years earlier, and just as quickly they had split apart.

From that small apartment in Sydney, the four underemployed software engineers had cobbled together a mapping demo that proved to Larry Page and Google what was possible in a web browser. Their proof of concept ended up inspiring a team at Google to leapfrog the competition. The Where2Tech team of four worked together on maps for only eighteen months, but their work became the basis of Google's mapping revolution.

Chapter 13

HELLO, GOOGLE EARTH

Google Maps was like a bolt of lightning striking Building 41. Google once again had demonstrated itself to be a place of unfettered, wild innovation—and shareholders took note. Google stock rose from $185 a share in February to $285 a share in June.

But while Google Maps had been an incredible step change for web-browser-based mapping, it still didn't rival the user experience possible with a dedicated app you download and install. It's much the same today: A mobile version of a website simply can't match the experience of a downloadable mobile app.

The fluid speed. The 3-D terrain and buildings. The fast, flythrough experiences. GIS tools like measurement, annotating maps, importing and exporting data sets: These were just a few of the features that only Keyhole EarthViewer had.

The upcoming Google version of EarthViewer was set to include a brand-new interface, a single Google search box (not multiple boxes dividing address searches from business searches), and new tools for annotation and measurement. It would enable fast access to *ten times* the satellite imagery data, at our highest resolution, and that data would be replicated out to thousands of servers in a half-dozen mega data centers around the world. Users in London, for example, would access data being served from the closest data center: Dublin, Ireland. And the new product was also going to be far cheaper.

While all of this work was completed through the spring of 2005, however, I still couldn't get us to agree on what we were going to call the thing.

One afternoon I was leaning back in a leather chair in John's office as we debated the name.

"I still think it should be Google Glll . . . GLOBE."

"You see!" John said, swinging around in his chair away from his monitors. He caught me. We were engaged in an ongoing debate—had been for the past two months—over what we were going to call the rebranded Keyhole EarthViewer: It was to be renamed as Google Something or Other.

The direction from Google marketing powers, including Doug Edwards and Christopher Escher, two of Google's preeminent marketing executives who had worked to craft the Google brand, was that any acquisition providing a service in line with the core mission of Google—of organizing the world's information and making it accessible—should be rebranded as a Google Something or Other. The Keyhole team was happy to oblige. We didn't want to be another Picasa or Blogger, Google acquisitions that had not changed their names.

As to the renaming of Keyhole's EarthViewer, two camps emerged.

"Google *Globe*!" I repeated, this time shouting it to John. He was firmly in the Google Earth camp, which I worried would limit us to environmental and earth science uses. Not that I was against earth science uses; I didn't want to limit the product by positioning it as a noncommercial tool for scientists. I hoped the product would find a foothold in everyday consumer uses like looking for a hotel, researching real estate, and choosing a restaurant.

For me and Google marketing director Doug Edwards, Google Globe also felt a little more whimsical and conjured up images of spinning classroom globes in elementary schools. A Googley name. It was a name we could underpromise and overdeliver on.

"Okay, say it ten times fast," John said. I fumbled again on the fourth *Google Globe*.

Truthfully, there was only one camp that mattered. John proposed Google Earth as the most simple and elegant expression of the original idea. After all, Earth had been the name of the application in Neal Stephenson's *Snow Crash* and Al Gore had worked with scientific luminaries to propose a digital Earth. Google Earth was the only name that John would accept.

Keyhole EarthViewer would soon be reborn as Google Earth.

At the time, I worked largely on four areas of the launch: naming the product, re-skinning the look and feel of the application, creating the website for the new product, and figuring out the pricing. (Keyhole EarthViewer was still $29.95 per year, less than half of its original price, and $400 for professional GIS users.)

I hired external interface designers, including trusted friends like Phil Melito, who transformed EarthViewer's interface from a professional, slick, and somewhat dark product to one that was more Google-appropriate: light, bright, airy, and fun.

Despite the fact that on day one I had agreed to the job title of product marketing manager, I was still fulfilling the role of product manager, too. More than once I recalled Bret's words that this would be impossible at Google. I was beginning to gain a firsthand understanding of this as we moved closer and closer to the Google Earth launch date. It was crazy. One afternoon in early May, I showed up in Building 40 at a crowded meeting filled with creative types, like designers and copywriters, with Google Earth interface designs in hand for review. I waited my turn for forty minutes to present the work and receive the approvals from various Google marketing execs. I was cut off on slide two. "Time-out," Debbie said, waving her hands above her head like a referee flagging me for an illegal hit. (If she'd had a whistle, she would have blown it.) "You're in the wrong meeting for this. You need to be in Marissa's product UI review meeting."

Was I a product manager or a product marketing manager? I didn't even know which meeting I should go to. The truth is, I was still doing both jobs. That might have been why I was working such long hours. It was challenging to be away from home so much, given that Shelley was pregnant with our second child and taking care of a petulant two-year-old.

And while I was working on the launch of Google Earth, I was also managing the marketing for the original Keyhole EarthViewer product. Plus, I was the marketing person overseeing Google Maps, which was exploding with interest from consumers and advertisers alike.

As part of the relaunch, we agreed to change drastically the pricing structure for the new Google Earth. For years, Keyhole had fine-tuned its product offering and its pricing to maximize revenue. Enterprise offerings sold to government agencies for hundreds of thousands of dollars. Professional licenses sold for $600 a year. Consumer versions sold for $79 a year. With the acquisition announcement, in October of 2004, we had cut those prices in half.

Now, in the summer of 2005, another price cut was planned.

Google Earth was to be introduced to the world with access to ten times the satellite imagery of Keyhole EarthViewer, thanks to Daniel Lederman's imagery acquisition deal with Digital Globe. There was an oversized storage hall in Building 41 filled with hard drives sent from Digital Globe for Chikai and Wayne's team to process, and a direct fiber-optic connection had recently been set up to transmit imagery. Google Earth was going to be much faster than Keyhole EarthViewer, as the database was now distributed across tens of thousands of servers, instead of a dozen. It would have more powerful tools, like annotation and measurement. It would come with the magic of Google search built into it. And the price?

Free.

Free? The idea was outrageous to the Keyhole team, but the direction from Larry and Sergey was clear. They would rather do something great for users and for the world than make an extra few (tens of) millions of dollars.

A palpable thrill moved through the team before the launch of Google Earth. There wasn't a more exciting place to be working, and there wasn't a more innovative product to be working on within the entire company. John was even developing a reputation as a star on the rise within Google. "Is it really going to be free?" said Ed Ruben to me one day with more than a bit of bewilderment that millions of people would soon be seeing his work.

We planned to continue to sell a professional version of Google Earth to enterprises and Keyhole's real estate customers. Google Earth Pro would continue to include better map-drawing and measurement tools, better printouts, and the ability to import in other GIS data sets, like Esri files.

With the pricing structure settled, I turned my attention to one last hurdle that stood in the way of mass adoption: our user registration process. While Google Earth was going to be free, John and Brian were adamant about still requiring users to register to use the product, and I knew that this stance would adversely affect mass-user adoption.

Traditionally on Keyhole's website, if a hundred people visited the site, only eight or nine of those hundred users would end up successfully downloading and installing EarthViewer. This was for a fourteen-day free trial offer with a required email registration. I had estimated that once the offer was for a free product (and not just a free trial), that maybe twice that number (16 to 18 percent) would install Google Earth.

By comparison, Google's photo software, Picasa, also a free service, was converting 35 percent of its website visitors into downloads and installs. Users were not required to register to use Picasa.

Brian was a fierce proponent of the registration process. He had been on the front lines of multiple spikes in traffic and Keyhole server outages. Despite going from eight to eight hundred servers, the service continued to go down frequently in response to Google home-page promotions and various news articles.

Brian was nervous about being able to keep the new Google Earth service up and operational: ten times the data, *and* branded

Google, *and* completely free. A user registration process was the last faucet at the end of a potential fire hose of demand. Now I was asking him to get rid of the faucet.

John was extremely reluctant to omit an authenticated email because he was worried about scraping of our valuable imagery data.

Despite Brian and John's resistance, I decided to take one last shot. I knew that our best meetings were always at the end of the day because people weren't rushed and had completed much of their work for the day. Late one Tuesday afternoon, I gathered a few critical decision makers, including John and Brian, to present one last pitch for a download process that would not require registration. I laid out the various scenarios, with projected numbers of users based on the Picasa experience. I also included Ed Ruben in the meeting, as it was Ed's infamous auth server system that controlled all access. Earlier, it had been the persnickety cash register; now it was the gushing faucet. It generated all license keys that would get mailed to users who registered.

I was lobbying for a download process similar to many modern-day mobile apps: Click an install button and open the app when it finishes downloading.

After I had presented the business case for not requiring registration, Ed's opinion shifted the overall sentiment of the room. He recommended that we eliminate the forced user registration. We could still build in the ability to not activate new users if the servers couldn't handle it, addressing Brian's chief concern by implementing a server-side kill switch.

"This is your decision, Brian," John said at the end of the

meeting. "Why don't you sleep on it and let us know what you think tomorrow morning?"

In a late-night email, Brian told Ed and me that he agreed with the position of no user registration. By early June of 2005, we were set to launch Google Earth. The product had been named: Google Earth. It was going to be *free*. The product had been updated with a friendly new Google interface. The features had been adjusted and refined, including adding Google search. All registration friction had been removed. The imagery database was ten times the size of Keyhole's. The database had been replicated across eight hundred servers globally. The press had been briefed by John. Google blog posts were written and approved.

The product had been presented to Larry. Thirty-five hundred T-shirts printed for the internal Google announcement at TGIF. Operations and support briefed and controlled by Lenette Posada Howard's team. Google legal approval complete. Server engineering approved based on forecast bandwidth and downloads. Product website created at earth.google.com and ready to go live.

Product website earth.google.com approved?

"So, you have this approved by Marissa and the UI review team?" Larry Schwimmer asked me. We were standing in the web-team area in Building 41.

"Yeah, I took Google Earth to Marissa's UI review team meeting two weeks ago. We are approved and we are set to go," I replied.

It was T minus twelve hours from the launch on Monday, June 27, and one of my final steps was teeing up the website earth.google.com, the landing page for Google Earth, for launch at midnight Eastern time, or nine o'clock Pacific time.

"No, I'm not asking about Google Earth, the software, I'm asking about the website. About *earth.google.com.* You've gotten written approval on this by Marissa, right?" Schwimmer asked.

Larry Schwimmer, employee No. 17, was known far and wide in the halls of Google. He was a quintessential Google engineer/ web developer. Super intelligent. Proudly geeky. Efficient. Shy, at times mischievous. And had a wry sense of humor. It had become part of the company culture that at every TGIF company-wide meeting, when Larry and Sergey ended their announcements to take questions, Larry Schwimmer was the first in line every week, often with a pointed yet wickedly funny question for Larry and Sergey. It had become kind of a company tradition that everyone took great pleasure in.

But Larry wasn't joking with me. "No public-facing website from any Google product goes live without the explicit approval of the UI review team and Marissa Mayer. None. No exceptions."

I had worked with the Google web development team. I had reviewed all copy with Michael Krantz, Google's lead copywriter. I had scoped and built earth.google.com as the welcome mat for those interested in Google Earth with web developer Jarod Lam.

A hundred percent of the press articles, blog posts, newsletter links, and Google home-page promotions were scheduled to point to one page—the Google Earth landing page—at the moment when the press embargo was lifted at exactly 9:00 P.M. (Pacific time). And Larry Schwimmer refused to push that one page live.

"Let me know if you get Marissa's approval," he said as he walked away.

I needed Marissa's approval. Even though she didn't control the product Google Earth, she did control all Google websites. All my emails and phone calls to Marissa had gone unanswered, though her admin told me that she was in her office.

I answered a phone call from John on my BlackBerry just as Schwimmer had left me hanging. He was in between interviews in New York City. Still wearing the walking boot from the basketball injury, John was hobbling around in New York City on a press tour with Eileen Rodriguez, a member of the Google PR team. Over multiple days, they had met with editors and reporters from *The Wall Street Journal, Newsweek, Time,* and many others. "Bill, you are just going to have to track her down and get her approval," he demanded. "Do you understand? Just go find her. *Now!*" It was rare for John to order me to do something. "Okay," I said. "I'm on it."

At 11:00 A.M., I rose slowly from the chair in Jarod's cube, where I would much rather have stayed, to make the long walk from Building 41 to Building 43, up the stairs, past the suspended Space X airship Larry had installed. I would just have to walk right into Marissa's office.

I shot past her admin, who had been purposefully ignoring me, sitting outside of Marissa's glass-walled office, just down from Larry and Sergey's mezzanine office. Before I uttered a word, I saw her surrounded by three product managers, studying her computer monitor, and five more Googlers waiting on the couch outside her door. "Marissa," I said. She was in midsentence with the product managers sitting with her. "Marissa," I repeated. She looked up at me incredulously. "I'm sorry to interrupt, but we are launching Google Earth tomorrow and I *have* to have your approval to go live with the earth.google.com landing page, and Larry Schwimmer will *not* push it live without your approval. I've been trying to reach you all morning. I've got to have your approval."

"Well, the UI review is on Thursday," she said curtly. "You and *John Hanke* will have to try to get on *that* meeting agenda and wait for approval until then just like every other PMM at the company." She waved her hand at the half a dozen or so PMMs waiting patiently outside her office. "That's the way the process works at Google." Clearly, she still saw me as a Keyhole outsider, and part of John's entourage.

"But we can't wait until Thursday. I need your approval *now,*" I argued. "We've got over eight press embargoes lifting tonight at nine. John is on a press tour with Eileen Rodriguez in New York. The Google blog post is teed up by PR. The product bits are distributed out across six data centers. Everything is ready to launch."

"What do you mean? *Ready for launch?* Well, obviously *not*!" she interrupted. "You do *not* have an approved website, so no, you are *not* ready to launch! Why didn't you bring this to the UI review meeting? You Keyhole guys are *always* doing this!" she added. "You guys *always* do this!" Marissa said in an even louder voice, slamming her hand on her desk for effect. The other product managers slowly edged away from the desk.

I said nothing. Probably because I wasn't breathing. I loved Google, and I didn't want to get fired on the spot. But I also didn't move. I just stood there blankly staring at her, not sure what to do. I didn't have a response because I didn't know what she was talking about. Was it possible that there was something about the Google Maps launch that she wasn't consulted about? Or maybe she was venting some lingering frustrations about John, who had usurped her control over the suddenly popular Google Geo product area?

Thankfully, after a few more tempered complaints about John and the Keyhole team, she gathered her composure. "Okay, let me look at it," she said, motioning me over.

I pulled open my laptop, took the seat of the product manager she had been meeting with when I barged in, and walked her through the Google Earth website. I knew it was good. Great, in fact, compared to other product landing pages. She approved it on the spot, and sent Schwimmer a quick note letting him know that we were good to launch.

"You have to promise, however, to come back on Thursday to present it to me and the UI review team," she said.

"Of course, I'll be glad to," I said.

Once Larry Schwimmer got the green light from Marissa, we were finally set to launch. I gathered with Larry and Jarod late on Monday night in their web-team area in Building 41. A fair amount of work was still being ironed out during this last-minute review. Even the smallest of changes (for example, an outline color of a button or a new Google Doodle on the home page) needed to hitch on to one of these hourly worldwide Google service updates. First the bits—the Google Earth software download file (about 28 megabytes)—were pushed out to thousands of Google servers around the world. Once these software bits were staged, we opened the front door at 9:00 P.M. (Pacific time). Google Earth was available for download at earth.google.com.

After testing that the earth.google.com site was working, I sent an email to the entire Google Geo team and headed home some time after ten o'clock. When I returned home, everyone was asleep. I slipped into bed and whispered to Shelley, "Well, we're live." I didn't know how Google Earth was going to be received, but I had successfully coordinated the launch of a Google product. Still riding the wave of adrenaline, I barely slept that night.

By the time I arrived the next morning, Ed and Brian and Chikai were huddled in Chikai's oversized cube in Building 41. Bright morning sunlight streamed through the nearby broad windows. The three of them observed the dashboards showing the bandwidth being sucked down on the thousands of Google Earth servers. The team fluctuated between states of wonderment and fear.

At 28 MB per install, Google Earth downloads and usage took a serious toll on the Google infrastructure. That first day Google Earth was downloaded 450,000 times. The mood in Building 41 was euphoric. Lots of high fives were exchanged that day. We were relieved to have made it through the launch. Ed, Brian, and Chikai continued to monitor the servers, looking forward to the moment when the traffic might subside and return to a more manageable load once the sun set on the East Coast. Traditionally, downloads and usage lightened when people turned in for the evening (often around nine o'clock or later).

But a curious thing happened. Instead of downloads decreasing, Google Earth downloads and usage continued to surge: Google Earth was going viral. Within the first twenty-eight hours, the Keyhole team had met its earn-out milestone of 500,000 downloads. It took twenty-eight hours to reach our two-year goal.

Google Earth also received glowing reviews from the press. Most mimicked the sentiment of *PCWorld* magazine, in which the headline read GOOGLE'S AMAZING EARTH, and the journalist went on to write: "It's a wonder we've got any work done. Google Earth is that spellbinding. It ranks among the best free downloads in the history of free downloads."

However, by Wednesday morning, Brian, Chikai, and Andrew were in full crisis mode. Late the night before they had met with Sergey and the server engineers because they were worried that the demand was beginning to threaten the performance of all of Google.com. That morning Brian met with John, who had just returned from New York City, and they made the decision to cut off downloads of Google Earth. Even Google, with its hundreds of thousands of servers, couldn't handle the load.

I scrambled with Jarod to throw together a web page with the following message: "We're sorry, but Earth is temporarily unavailable." As a special marketing touch, I asked Jarod to include a satellite photo of the Earth at night on the page. Google Earth had gone dark. That day I skipped most of my meetings, but kept Debbie and Marissa in the loop of what was happening. A pattern of going up and down persisted throughout the weekend. Andrew and server engineers distributed Google Earth across even more servers and the downloads were turned back on. Then the downloads were switched off once more under a crush of hundreds of thousands of new users each day.

On the seventh day after the launch, Brian, Chikai, and Andrew finally had worked through a solution that could support the demand, including distributing Google Earth out over several more data centers and thousands of additional servers. Still, demand for Google Earth did not subside like typical software launches. Instead, it was downloaded 10 million times during the first month, and it continued at the high daily download rate for months. We averaged 300,000 to 500,000 new users daily, sometimes more. One day in October, Google Earth was installed over 900,000 times. I recalled our initial meeting with Larry and Sergey the year before, and the ridiculous question that I posed about 10 million users or $10 million during the first year. We surpassed all of these possible expectations within the *first* month—and I was beginning to understand firsthand what Larry meant by "thinking bigger."

In early February, Google had no mapping products. By the end of June, we had launched arguably two of the most transformative tools for interacting with and navigating our world. Overnight, millions of people switched to Google Maps and Google Earth for finding their way in the world. I'm not sure we were ready for the responsibility that came with this widespread popularity. The inviolable trust in the Google brand—built up and guarded carefully by the Google founders over the past five years—now projected a new expectation onto John's Google Geo team. Our millions of users assumed, if not demanded, that we get maps right, and the satellite imagery added an even greater belief in what we put forth as geographic truth. And we usually did get it right. But not always.

One leisurely Friday afternoon, after lunch, I was catching up with Dede about our weekend plans. Dede's cube was just outside John's office and faced toward the expansive windows overlooking Shoreline Amphitheatre. As we talked, I heard something.

"Do you hear that?" I said to Dede.

"Maybe there's a concert happening today," she said, swiveling her chair to look out the window. Sometimes late in the day we would hear a band warming up for a performance that evening. But it was too early for a concert rehearsal. As we stared out the window, the noise grew louder. And then I began to see them.

Protesters. Protesters by the dozens. No, protesters by the hundreds. And they were being followed up the sidewalk by multiple news crews. Google security monitored the burgeoning crowd. A man with a megaphone in his hand led the protesters. The Northern California sun shone brightly. "What are they protesting?" I wondered aloud.

"Well, that sign has a Google Maps icon on it," Dede commented.

Multiple posters with handwritten slogans dotted the loud but orderly crowd outside the office windows. "PRC." "Taiwan Is a Sovereign State." "Shame on Google Maps & Google Earth!" "Taiwan Is Not China." "Let Taiwan Be Taiwan." The protesters stopped on the sidewalk outside Building 41, maybe thirty yards from John's office. Two television news crews—one from San Jose and one from San Francisco—set up shop.

John was still at lunch; I had just left him with Michael, Brian, and Daniel at Charlie's Café. I decided to give him a call.

"What's up?" he asked, picking up immediately.

"Hey, just wanted to let you know that your one o'clock protest is here," I said, looking out the window at the gathering masses.

"What?"

"Your protest. The one o'clock protest. They are here. Right outside your office. Should I tell them to wait until you finish your lunch?" The chants grew louder. "Can you hear them?" I held my phone up to the window. Dede rolled her eyes, shook her head, and chuckled.

This border dispute was the first of many. The Taiwanese, as represented by the protesters outside our office, saw themselves as completely independent of China. Taiwan sees itself as its own sovereign country, while China considers Taiwan a province of China. China lays claim to Taiwan, referring to the territory as "Taiwan, PRC," or "Taiwan, People's Republic of China."

When we launched Google Maps that included Taiwan, by labeling the territory *Taiwan, PRC*, we had effectively taken sides in a heated international dispute. The foreign ministry of Taiwan and legislators from the pro-independence Taiwan Solidarity Union were urging Taiwanese in the United States to protest Google and demand that we change the name.

We were in a real pickle. Changing the name would create a new problem: Now it would be the Chinese government that we might provoke—and Google's relationship with the Chinese government was already very complicated. It should come as no surprise that a company whose mission is to make information accessible would have a challenging time operating in a country whose mission is to control and censor information.

Another dispute, between Nicaragua and Costa Rica, generated international headlines, claiming that Google Maps had nearly started a war in Central America. With one data push by Wayne, our team drew the border between the two countries just to the south of what was generally accepted as the geographical division along the Río San Juan boundary. As a result, it looked like Google had unknowingly handed over a few square miles of Costa Rica to Nicaragua.

Like his Taiwanese counterparts, the Costa Rican foreign minister complained to Google, and we corrected the line in a database update some months later. But during those intervening months when we seemingly "annexed" the land to Nicaragua, that country situated fifty soldiers on an island called Isla Portillos and claimed it as its domain, saying, effectively, "Google Maps says that it is our territory."

Our mistake was not completely arbitrary: There was a reason our border database had included this geographic variance. It turns out that the territory had been disputed in the 1800s, with Costa Rican coffee planters voting to secede from Nicaragua in two border towns in 1824. There were seven treaties drawn up, but none were ratified by both countries. The border dispute was finally settled by a treaty in 1858, but the river remained hotly contested, as it was a potential location for a canal that might provide passage between the Atlantic and Pacific one day.

This border had been so controversial that in 1888 President Grover Cleveland dispatched a surveyor as an arbiter to clarify the treaty. Complicating matters, the Río San Juan has naturally drifted to the north, into Nicaragua, over the years. Nicaragua has dredged the original channel in order to maintain the official boundary. We had drawn our line on top of this complicated backdrop—and reignited a centuries-old border dispute.

As Google Maps fanned out, we were drawing lots of lines. Daniel's team quickly licensed data from dozens of different providers and utilized some of our data sources from the days of Keyhole, too. Google Maps and Earth were on their way to becoming the most comprehensive map set, with the largest map viewership on the planet. And the planet expected Google to get it right.

It wasn't always easy.

We appeared to annex a shipping channel in a place called Dollart Bay, which separates the Netherlands from Germany, giving over an important maritime port of entry to the Dutch, much to the chagrin of the German port city of Emden (home to a Volkswagen exporting automobile plant). We faced animosity from both sides around the proper naming of Londonderry/Derry in Northern Ireland. South Korea took exception with Google Maps for labeling what it refers to as the East Sea with the label that is claimed by Japan (Sea of Japan). Palestinians created an online petition with over twenty-six thousand signatures in response to Google's labeling of the West Bank and Gaza Strip. We faced scrutiny in the Ukraine and Russia over ownership of the Crimea peninsula. In India, a law was drafted about how borders with Pakistan were to be drawn, with the penalty of jail time for the mapmakers who did not abide by India's view of ownership of the Kashmir region.

All told, over thirty countries were actively involved in border disputes at any given time, some of which had been contested for hundreds of years and had changed dozens of times. The Spratly Islands, in the South China Sea, are currently claimed by five different countries.

As the new leader of all mapping efforts at Google, John was responsible for overseeing and resolving these issues. He was forced to become an expert: the final arbiter of dozens of international naming and border disputes. Fortuitously, his years at the State Department right after college served him and the company well. John and the mapping team developed a thoroughly researched, documented, and published methodology. A recommendation was formulated, often by working with Google country managers on the ground in the territory in question, and John made the final decision about how Google's mapping products would handle the dispute. For boundary disputes, both borders were drawn as yellow dashed lines (instead of the usual solid line). For naming disagreements, we defaulted to the United Nations as our guide and sometimes included both names. Andria Ruben, who had sold Keyhole software to the United Nations during the Keyhole days, helped to define this solution.

In some instances, users viewed different versions of Google Maps, depending on *where you accessed Google Maps from.* (This isn't something widely publicized.) For example, a user in Japan would be shown a Google Map where the sea to the west of Japan was called the Sea of Japan. A user in Korea would be shown a Google Map where that very same body of water was labeled the East Sea.

On that sunny Friday afternoon, it was eye-opening to witness the passion of those Taiwanese protesters. For me, it was the first time I realized that I was a part of something bigger. That this whole Google Maps and Google Earth thing was more than another consumer web service.

People loved Google and our new mapping products. They trusted Google. They wanted us to get it right. No, they demanded we get it right.

Chapter 14

DASHBOARD FOR THE PLANET

The launch of Google Earth could have been the final chapter. That moment—and all of the work that led up to it—went far beyond the founders' original dreams and vision when they started Keyhole. Only two years earlier, the company was nearing the end of the runway, mere weeks from going out of business completely—and now we had millions of users spanning the globe using our product in unexpected and sometimes humbling ways. This could have been the end of the story, but in many ways it was just beginning.

Hurricane Katrina slammed into the coastal regions of southeastern Louisiana on Monday, August 29, 2005. Shelley (six months pregnant at the time) and I monitored the storm closely, as we were scheduled to fly into New Orleans the following weekend before driving to the pristine beaches of the Florida Panhandle for Labor Day weekend with friends. The weather system churned into a Category 5 storm in the Gulf of Mexico and made landfall as a Category 3. Despite the massive storm surge, New Orleans appeared to be spared. On Monday evening, our getaway with friends was still on.

On Tuesday morning I received a call from John Rareshide, a longtime friend of mine and John Hanke's from our UT days. A New Orleans native and resident, he had evacuated to his in-laws' house in Beaumont, Texas. "Turn on CNN," he said. "The levees have broken and the city is filling up like a bowl of cereal. We're fucked. The whole city is fucked." I turned on CNN to see a reporter standing in the French Quarter as the waters began to rise up his legs. I was surprised and alarmed to see a handful of people on Bourbon Street who hadn't evacuated yet.

"I need you to get John to fly the Google satellites over my house," Rareshide said. "We don't have our own satellites," I told him. I thought to myself, *Does the world think we have our own satellites? What a crazy idea. Google having its own satellites.*

Throughout that morning, the conditions in New Orleans grew increasingly dire. With the sudden popularity of Google Maps and Google Earth, my friend John Rareshide wasn't alone in calling for Google to do something, anything, to help with the situation. As CNN continued to use Google Earth on the air to cover the floodwaters, these calls became more forceful and desperate. Couldn't we do something to get updated imagery of New Orleans?

By that evening, John Hanke had determined that there was something we could do, and convened an impromptu meeting on the couch outside his office with Chikai, Wayne, Brian, and me. He had been in contact with an aerial imagery pilot out of St. Louis named Kevin Reece. NASA and the National Oceanic and Atmospheric Administration (NOAA) had contracted with Reece and his team to fly over New Orleans and capture updated aerial imagery of the flooded city. John asked Chikai and

Wayne to figure out a way of establishing a pipeline of this new data so that Reece's updated imagery could make its way into Google Maps and Earth. (It was actually quicker to publish this data first as individual KML image overlays in Google Earth.) Wayne and many on the team worked until three o'clock in the morning to process and update Google Maps and Google Earth. This high-resolution imagery update provided the evacuated residents of New Orleans—who were prohibited from reentering for weeks—their first look at the devastated state of their city and their homes.

About ten days later, a call came into the main Google number, which was then forwarded to me. Because I wasn't at my desk, the caller left a voice mail. It was from Ron Schroeder, a staff sergeant of the medevac unit with the Coast Guard helicopter fleet operating in New Orleans. The message almost made me drop the phone. I turned to my office mate, Daniel, and said, "Hey, you've gotta hear this." I replayed the voice mail on speakerphone. Schroeder's message said that the Coast Guard was using Google Earth to save hundreds of lives in New Orleans.

Immediately, I called him back. Schroeder explained how Google Earth, which had only been released two months earlier, provided a critical link between despondent residents trapped on New Orleans rooftops and attics and the dozens of Coast Guard rescue helicopters circling the city. This is how it worked: Callers to 911 were patched through to the Coast Guard radio dispatchers. The callers were asked to identify their exact locations, and of course the individuals responded with a street address and likely a cross street. For example, "4130 Toledano; it's two blocks south of St. Charles on Toledano."

This presented a challenge for the Coast Guard navigation systems, which weren't set up for street addresses. Members and boats of this maritime branch are almost universally deployed somewhere *off* the coast: to capsized boats, cargo ships with questionable shipments, and offshore oil rigs. In other words, locations that don't have street addresses. Before Katrina, the Coast Guard only received distress signals that included the latitude and longitude information, and those coordinates were used by the helicopter pilot to fly to the location in distress. Can you imagine calling 911 and having the dispatcher ask for your latitude and longitude? That's exactly what was happening, Schroeder explained, until they downloaded Google Earth.

With Google Earth, the Coast Guard helicopter dispatcher could enter in a street address, like 4130 Toledano. Then, by simply rolling over the house with his mouse, he could get the exact latitude and longitude of the rooftop and then radio this information up to the nearest helicopter available for a rescue. It is a common and useful feature of Google Earth: Rolling your mouse anywhere on the screen returns the latitude, longitude, and elevation of the exact spot where you place the cursor.

The caller could also explain what part of the roof they were on, or if they were on a detached garage or other structure on the property. This way, the dispatcher could obtain the precise latitude and longitude, down to one-meter accuracy, of where the helicopter needed to fly. Thanks to the updated imagery being piped in from Kevin to Wayne, the dispatcher was able to radio to the pilot additional situational information, such as fallen trees and downed power lines, among other likely obstructions.

"You guys are absolutely saving lives down here, and I think this story needs to be told. You should be really proud about this," Schroeder told me. "Of the thousands of rescues we've done down here, I'd conservatively estimate that over four hundred lives have been saved as a direct result of Google Earth." I can honestly say that I had never before—and have never since—felt so proud to be a part of a team.

Schroeder was extremely thankful for the software and especially grateful for the work we had done to update the imagery of New Orleans. We had updated this imagery without even knowing of Schroeder's operation and use of Google Earth. He went further, telling me, "I think this is a story that should be told, and I'd be happy to help Google get the word out." I thanked him for his extraordinary work and promised to get back to him.

Immediately, I walked over to the office of David Krane (head of Google PR). I told David and Megan Quinn, another member of the PR team, the story and played the voice mail for them. I asked them what they thought. Did we want to reach out to Schroeder and place this story with a reporter?

Quinn and Krane looked at each other. "No, we wouldn't do that," Krane said. "As amazing as the story is, Bill, it's just not who we are. It's not who Larry and Sergey are. We as a company don't like to beat our chest about something that we've done. It's not our style to pat ourselves on the back for something like this." Megan agreed.

Like many other things at Google, it flew in the face of every marketing lesson I had previously learned. In other marketing roles, I was used to looking for every potential morsel of goodness to take credit for and then broadcast in a public setting. Now here I was, a part of a product that had reportedly saved hundreds of lives, and I was being told, "No, actually, we don't want to talk about this publicly."

When I called him back to decline the offer, Schroeder was as stupefied as I was. But my bet is that, like me, he came away from this experience with a little more appreciation for Google and Google Earth.

Krane and Quinn did allow some exposure—although only internally—for this voice mail. With a smile on his face, Larry played the voice mail for the whole company at the next week's TGIF, and it was met with an enthusiastic ovation. There was no press article or blog post about what happened, but I think everyone at Google left that TGIF with a lot of pride in the company they worked for. I know I did.

Hurricane Katrina was the first time that Google Maps and Earth responded to a natural disaster. After this coordinated effort, a Google crisis response team was established with special maps, using the Google Maps API, to aid and support communities and cities during natural disasters. This tool, updated with new imagery, was also embedded in multiple media websites and the State Department in order to inform wider audiences. During the Haiti earthquake in 2010, a group of volunteer engineers developed a Google Maps API app, a sort of people finder, for individuals looking for each other (even if they no longer had cellular connectivity). This tool provided a registry and message board so survivors could post and search for specific details about one another's status and exact whereabouts. (This app was launched in English, French, and Haitian Creole three days after the devastating earthquake.)

Disaster relief was the first of many new, unexpected uses of

Google Earth. In October, John forwarded an email to me saying, "Can you please meet with this woman?" The request for a meeting had come from Rebecca Moore. She was an environmental activist who was using Google Earth to fight high-powered, deep-pocketed logging operations in the Santa Cruz Mountains.

After only three months of launch, we had already heard about dozens of different environmental groups embracing Google Earth to visualize a variety of environmental causes, from visualizing rates of deforestation in the Amazon rain forest to detecting and mapping out illegal commercial fishing activities. Google Earth represented an opportunity to change radically the way that environmental monitoring was done and to effectively crowd-source the efforts to keep an eye on our planet. Rebecca Moore's fight against the proposed logging operations in the Santa Cruz Mountains was right in our backyard: about one hour away from Google.

I thought the meeting would entail a casual cup of coffee with a friendly do-gooder who maybe wanted us to sponsor a fundraiser. I had given the meeting little thought and hadn't researched Moore's background.

I decided to meet with Moore on the couches outside John's office, with the hopes that they could say a quick hello in passing. A computer science and data mining professional educated at Brown and Stanford, Rebecca Moore was not who I expected. A no-nonsense woman with shoulder-length blond hair and a Northern California spirit, Moore was dressed in jeans, a blazer, and Birkenstocks. On her laptop, she took me through a Google Earth–based KML presentation that she had used to stop the logging of a thousand acres of redwood trees in the Santa Cruz Mountains. (Though the product name was Google Earth, Brian decided to keep the data standard labeled KML in homage to Keyhole.)

She showed me the pathetic little public-notice flyer that had been sent by the logging company to the two thousand residents of her community, with a tiny black-and-white map created to minimize the perceived impact of the proposed logging operation. The image brought to mind a book I had read by Mark Monmonier titled *How to Lie with Maps,* which highlighted the ways that maps can be used to selectively redraw reality.

When Moore redrew the proposed logging operation in Google Earth as a KML file one weekend in August of 2005, the reality of what was being proposed was shocking: The swath of redwoods being logged was close to six miles long, and the operation even went right up to the local day-care center. That wasn't exactly highlighted on the public notice the community had been sent. For dramatic effect, Moore even created 3-D models of log-carrying helicopters that would likely be circling the community. Thanks to Moore's Google Earth presentations, the proposed project was summarily rejected during a recent town hall meeting.

I was stunned by what Moore had done, but knew that Google wouldn't want to be involved in any additional press attention related to Moore's win. When I asked her how Google could be of further help, Moore surprised me once again. She didn't want Google to participate in some sort of press outreach, nor did she ask for a $2,500 check so Google could be listed as a sponsor of her group's annual banquet. Instead, she launched into a vision of what could be done with Google Earth to essentially save the planet. As our conversation entered its third hour, it became clear to me that Moore was serious about the matter and knew way more about the subject than me or John or Brian.

She explained the primary missing elements in the Google Earth data-processing workflow. Namely, we needed to create a tool or a website by means of which environmental and earth science groups could upload their data in a structured way for redisplay as a Google Earth layer. The tool would process and host the data, off-loading two of the technical hurdles standing in the way of the visualization of all kinds of environmental data in Google Earth. "I don't know if you guys realize what you have done," Moore said toward the end of our meeting. "Do you realize that you have created the dream tool for the environmental community? There's so much excitement out there right now, thanks to Google Earth."

Later that day John asked me about the meeting.

"Here's the deal," I said. "You are going to have to hire her. Like immediately." And John did. Rebecca Moore would report to Brian and was placed in charge of the Google Earth Outreach Program, leading the use of Google Earth to map out a variety of environmental issues. As of the writing of this book, Moore is still at Google, working with her counterpart Jenifer Austin Foulkes to create a "living, breathing dashboard for the planet," as she calls the next generation of Google Earth.

Environmental groups weren't the only organizations to see the value of Google Earth. Groups like the United Nations and the U.S. government were already customers of Keyhole, and now there were millions of new potential viewers of the data layers that they might create, just as Moore predicted.

Even the then-president of the United States, George W. Bush, was a user. In an interview on CNBC, when asked if he used Google, he said that he liked to use "the Google" to fly over his ranch. Later he commented on the work that had been done by a group called Bright Earth to shine a spotlight on the burned-out villages in Darfur, Sudan. "I also saw an interesting new venture [*sic*] arranged with Google Earth. As a result of this partnership, millions of Internet users around the world can zoom in and see satellite images of the burned-out villages and mosques and schools," Bush said. "No one who sees these pictures can doubt that genocide is the only word for what is happening in Darfur—and that we have a moral obligation to stop it." In 2007, the international community responded by deploying twenty-six thousand peacekeeping troops to the region to stop the violence and guarantee safe distribution of humanitarian aid.

These unexpected uses of Google Earth also included countless new discoveries: new species of animals, new islands, a new barrier reef, a miniaturized military training ground in China, migrating cow deer in the Czech Republic (that naturally aligned themselves according to the Earth's magnetic poles), and a variety of others.

One newspaper story that made the rounds at Google was about an Indian man named Saroo Brierley who used Google Earth to find his way back to his hometown in India, retracing his steps from his accidental train trip across India at age five. His reunion with his mother via Google Earth was a feel-good story for the team. In 2016, Paramount Pictures released an Academy Award–winning movie, *Lion,* based on Brierley's best-selling memoir, *A Long Way Home.*

Of all of the surprising uses of Google Earth, the initiative that grabbed John's attention the most was the efforts to map out the ocean floor. John was invited to Madrid to present Google Earth at the National Geographic Society of Spain's annual conference. After his presentation, Dr. Sylvia Earle got up to deliver her talk. Earle is an oceanographer, an aquanaut, and a pioneer in the field of marine biology and underwater exploration. During her research and work, she has logged weeks living in underwater habitats. In 1979, Earle set a world record by diving 1,250 feet to the sea floor near Oahu. She later served as the first female chief scientist at NOAA and was honored by *Time* as the first Hero for the Planet. Since 1998, Earle has served as a *National Geographic* explorer in residence.

As she started her presentation, Earle thanked John for his demo of Google Earth, but then challenged him in front of the audience. "But, John, shouldn't you really call it Google Dirt?" she asked. "After all, you still can't explore over two thirds of the Earth in Google Earth." While the sharing of data using KML files had proven to be extremely useful to the earth sciences community, oceanographers had been vocal in their frustration that Google Earth didn't allow users and researchers to explore below sea level. John agreed with Earle; she certainly had a point.

From my conversations with John, it was evident that Earle struck a nerve. With Larry Page's support, and Rebecca Moore and Jenifer Austin Foulkes's stewardship, a new mapping project was born. Created in 2011, Google Ocean is an initiative to map the ocean floor, allowing Google Earth users to dive below sea level to explore underwater. Google Ocean ended up broadening the reach of Earle's efforts to create marine protected areas, or Hope Spots, as she calls them. Earle's organization Mission Blue aims to set aside and protect these Hope Spots, much like national parks, on the bottom of the ocean. She has used Google Earth, along with Google Ocean data, to help create fifty official designated and protected areas on the ocean's floor. You can view them today either with a pressurized atmospheric diving suit or as a KML layer in Google Earth.

In addition to these far-reaching environmental efforts, an eclectic mix of celebrities and politicians paid visits to Building 41. Musician Peter Gabriel met with John to discuss a project to create ambient sounds from all over the Earth and display them as a showcased layer on Google Earth. Actor Woody Harrelson came to talk to us about strip mining in his home state of West Virginia. Colin Powell came to Google and I was excited to demo Google Earth for him, but something more exciting happened that kept me away—the birth of my second child, Camille. (Isabel was three years old by this time.) Michael Jones stepped in for me. Bono asked us to work on a map-based animation for U2's Vertigo Tour. We were happy to oblige, and John and I felt the need to personally inspect the project when they performed in Oakland.

Former Vice President Al Gore came for a demo one Friday, and stayed for the company-wide TGIF gathering later that afternoon. I was standing with John and Daniel in my usual spot toward the rear of Charlie's Café, with a cold beer in hand, listening to Larry and Sergey roll through the news from the week. Behind me, I heard a bit of commotion. A small entourage, along with a few members of our PR team, escorted Gore into the cafeteria. He stood right beside me, and I whispered a quiet hello.

He is likely the only non-Googler I had ever seen at the weekly TGIF; access is strictly limited to full-time employees. But for Gore, who was on the advisory board for Google, an exception had been made. *So, cool, right? Al Gore is standing right next to me.*

Cool except for the cosmically awkward coincidence of Larry Page's very next slide.

It was common practice at TGIF for Larry to introduce any key new hires that had happened during the week. Up on the screen, he flashed an image of someone who was joining Google in the role of chief technology evangelist: Vint Cerf.

Do you know what Vint Cerf invented? Ever heard of something called *the Internet*?

Yeah, that's right. In the mid-1970s, while working at the Defense Advanced Research Projects Agency (DARPA), Cerf codesigned a protocol by which packets of data could be disassembled into smaller packets, transferred over a network, and then magically reassembled at its destination—an IP address. Transmission control protocol/Internet protocol (TCP/IP) is the foundation of how all data still flows today.

For good measure, just to make the situation more awkward, Larry flashed the next slide with Cerf's picture and the phrase *Father of the Internet,* which is how he is widely regarded in engineering circles. I immediately let out a Texas-size guffaw, elbowed Gore in the ribs, and shouted out, "Hey, dude, I thought that was you!"

No, I didn't do that. But if I had wanted to be fired on the spot and quickly escorted out of the building, that would have been one hell of a way to go. Instead I kept my mouth shut and held my breath, as did all the other Googlers in the immediate vicinity, and was quite happy when Larry moved on to the next slide.

I should point out that months later I had the chance to sit in on a tech talk given by Cerf, and he was grateful for the crucial role that Gore played in the commercialization of the Internet, graciously saying that Gore, during his years as a U.S. senator from Tennessee, in fact had been instrumental in opening the network up for nonmilitary purposes.

The doors were also thrown wide open for an array of movie, television, and sporting event tie-ins for Google Earth, too. Google Earth made its way into *The Bourne Identity*. It was used on the television shows *24* and *CSI*. It was utilized to cover cycling events, like the Tour of California and the Tour de France, thanks to a Googler named Dylan Casey (who competed as part of Lance Armstrong's U.S. Postal Service team). It was used by NBC as part of its Winter Olympics coverage in Torino, Italy.

On my office whiteboard, I kept an ever-expanding list of Google advertisers and partners that wanted to do some sort of joint marketing promotion: Target (the company started painting store rooftops with its logo), Disneyland, Audi, Virgin America, Dell Computers, Mazda, PepsiCo, Expedia, Sierra Club, Fiat, JetBlue, Sony, *Pirates of the Caribbean*, and Santa Claus (NOAA wanted to create a Santa Tracker). Plus, it seemed that every sales rep at Google had at least one advertiser interested in doing some special promotion featuring a Google Maps and Earth mash-up.

Of all the partnerships and entertainment tie-ins for Google Earth, one in particular stands out for me. The inquiry came in to Jonathan Rosenberg, who was an alum of Claremont McKenna

College in Southern California. Another alum, John Corey, had emailed Rosenberg with a question: "Is there anyone on the Google Earth team that might help me create a movie animation that I am working on for the Jumbotron display at the Rose Bowl in Pasadena?"

"Hey, Bill, come in here," John said to me one early December evening. "Shut the door."

I did what he said, more than a little worried, but the restrained smirk on John's face gave me hope. "I want you to read something," he said, moving from his chair so I could take his seat.

"Oh my God!" After reading the email, I jumped up and we high-fived each other.

"I'm going to forward this to you, but you've gotta play it cool, okay?" he said.

For over a week, John and I had been looking for tickets to that year's Rose Bowl, an epic matchup that we had been dreaming about our whole lives (practically). For the first time in thirty-five years, our Texas Longhorns were playing for a national championship against the undefeated USC Trojans, who were looking to win their third consecutive NCAA college football national championship.

The following day I emailed Corey, telling him that I would be happy to help him on the Google Earth movie, signing the email with the not-so-subtle hint: "Bill Kilday, University of Texas class of 1990."

Later that week, we met in my office to discuss the project. To be honest, I did consider myself a pro at using the movie-making feature of Google Earth. After all, I had sat with software engineer Francois Bailly as he built the feature, giving him feedback about how the tool needed to work for marketing types like me. Corey was happy to have my help. His concept was to create a video for each team that would include highlights of every game they had played, with a Google Earth animation flying the viewer to each of the stadiums in which the games had been played. On the way out of Building 41, Corey turned to me and said, "So you went to UT, I see. Do you have any interest in going to the game?"

"Well, yeah, now that you mention it. I would like to go to the game." I casually mentioned that my boss John Hanke was also a University of Texas alum.

"Let me see what I can do," Corey responded.

The email that came in from Corey the next morning set off another round of disbelief. Corey offered up four VIP passes to the game, which meant that in addition to our fifty-yard-line seats, we would also be allowed onto the field before and after the game. We invited Dan Clancy and one other Longhorn along for the trip. (Rosenberg was invited, but kindly declined, suggesting we find another Texas alum to take along.)

As John and I stood on the field before the game, watching the Google Earth animation loop in front of over one hundred thousand fans, he typed out an email on his BlackBerry to Jonathan Rosenberg, Debbie, and other marketing executives, praising me for the extraordinary free promotion I had orchestrated. (Note: John did not include a photo, as his leading-edge BlackBerry did not take photos.) Debbie forwarded his email to the entire Google marketing organization.

It is not an exaggeration to say that the 2006 UT-USC national championship Rose Bowl game is widely considered to be one of the most exciting college football games in history. The Longhorns won a thrilling, improbable, come-from-behind 41–38 victory against the two-time defending champion on Vince Young's mythic run on fourth down from the nine-yard line with nineteen seconds to play.

As the clock hit zero, John and I sprinted out of the stands and onto the frenzied field like giddy schoolkids, high-fiving players and Longhorn luminaries like Michael Dell, Matthew McConaughey, and Lance Armstrong. I locked arms with a group of offensive linemen in front of the Longhorn Band to sing "The Eyes of Texas," and then I caught John running around the end zone and taking pictures with his camera. John was snapping photos of the Google Earth animation that was looping on the Jumbotron again as the Rose Bowl crowd roared in delight. The falling confetti might as well have been for John and me.

It was a storybook ending to 2005, a remarkable year in which somehow I found myself a part of two of the most successful product launches in history: Google Maps and Google Earth. Both had exploded in popularity, both had millions of users, and both had helped drive the price of a single Google share from $200 at the beginning of January to $436 at the end of December. My good friend was now firmly established as the leader of both products, and all signs pointed to expanded investment in Google's Geo division: I was soon to be getting a team of marketing people to work with Ritee and me.

I soaked it all in—the confetti, the trophy ceremony, the looping Google Earth animation—though I had to remind myself of the Lyle Lovett song, "All Downhill from Here." It had been one hell of a run, but by the end of 2005, I was beginning to wonder how much longer I could keep up the pace. I had been redlining at work for many months, and, with two baby girls at home, I found it was starting to take its toll. I had also run into lots of friends from Austin on that trip, and I found myself wanting to go home with the armadillo.

How much longer could I keep up with John Vincent Hanke? I knew him well enough to know. Yes, John might have been momentarily content as we stood there on the field of victory. But as soon as we woke up the next morning, he'd be back at a full sprint.

Chapter 15

YOU: A BLUE DOT

Dozens of themed slot machines—*Wheel of Fortune! Triple Diamond Payout! The Price Is Right!*—called out to me as I walked with the excited masses of arriving travelers through Las Vegas's McCarran International Airport. The departing, bleary-eyed people leaving Las Vegas shuffled through security like a herd of cattle. On that bright, cold Thursday morning, outside of baggage claim, the cab line was at least an hour and fifteen minutes long. A banner hung over the throngs of arriving people: *Welcome to the 2006 Consumer Electronics Show!* The 170,000 attendees of the Consumer Electronics Show (CES) were mostly well-dressed Asian and German businessmen on expense accounts.

At CES, electronic manufacturers connect with potential distributors and retailers to announce products and secure retail and digital shelf space. Many bottles of sake, vodka, and whiskey consumed over the next four days would help seal the deals pursuing a slice of the $213 billion U.S. electronics market. I stood out amid the business-attired crowd in my bright-green long-sleeved Google shirt, which was a part of my booth-duty uniform. It was January 5, 2006, the day after UT's thrilling championship victory over the USC Trojans, and I had traveled from Pasadena to Las Vegas to work the Google Maps and Google Earth demo stations with Ritee at the show.

If I had still been working at Keyhole, I would have had to miss the game and have arrived in Vegas at least two days earlier, carrying computers, monitors, and swag onto the trade-show floor. CES offers about 1.85 million square feet of booth space, which is the equivalent of 393 basketball courts. The largest booths for companies like Samsung, LG, and Microsoft take eighteen days to construct. Back in the Keyhole days, I would have had to outfit a ten-by-ten-foot booth, taping down power strips, setting up Wi-Fi routers and networking cables, and begging show organizers for better booth spots. Now with Google, I simply rolled onto the CES floor, finding my way to the forty-thousand-square-foot, yellow-carpeted, Lego-themed booth space with the massive illuminated Google weather balloon floating above it. Everything was already set up for me, with a Google Maps demo station loaded on a four-foot green Lego piece and a Google Earth demo station arranged on a four-foot blue Lego. In fact, all Google product demo stations were displayed on Legos, and multicolored beanbag chairs were scattered throughout the booth/lounge.

I stored my backpack in a secret compartment underneath the Lego table and got to work, giving demos and answering questions. Countless manufacturer representatives stopped by to show me their new connected handheld devices, all with the desire to embed a version of Google Maps inside them. Many of these demos ended with an inquiry about how to get a job at Google, either for themselves or for a very smart college-aged son or daughter. Most attendees had never met someone who worked at the six-year-old company with five thousand employees.

It was the first (and only) time that Google would have a booth at CES. Our presence was largely the result of the fact that Larry Page was giving the keynote address for the conference the next morning. I knew Larry's keynote well; several segments of his presentation were mapping related, and I had worked with the Google PR team, reviewing sections of his speech.

The CES keynote is considered to be a milestone moment on the technology world calendar; the entire electronics industry and technology media train their attention on this single presentation as a kickoff for the year and a chance to hear about the future of innovation. Larry's speech was highly choreographed and well rehearsed. Google wanted to avoid a repeat of Bill Gates's 2005 keynote when an introduction of a new Windows Media Center product crashed in a blue-screen-of-death debacle.

At nine o'clock the next morning, the lights dimmed in the five-thousand-person ballroom at the Las Vegas Convention Center. Even the overflow rooms were packed, with attendees waiting to find a spot. I saw the long lines to get in and decided to watch a closed-circuit video feed piped into our booth on the trade-show floor. A thumping soundtrack pulsed in the darkness. Google Earth—with an animation showing worldwide Google search traffic—was projected onto the stage and the dozens of large screens throughout the room and overflow areas.

To balance out Larry's deliberate speaking style, the presentation also would include cameos from comedian Robin Williams (who delivered a very politically incorrect vision of a future embedded Google Brain complete with fax machines to, ahem, output the search results; this section was blacked out of the feeds due to his off-color humor) and NBA star Kenny Smith. Larry had no clue who the two-time NBA champion was, but knew his first guest onstage very well: Stanley.

As the lights came up, Larry rolled onto the stage, standing on the bumper of a customized blue-and-red Volkswagen Touareg—Stanley, the Stanford-created, self-driving, off-road SUV that had recently won the 2005 DARPA Grand Challenge. Larry had witnessed Stanley's victory of the 132-mile race firsthand, with over a hundred sharp right and left turns, through the Mojave Desert. Three months earlier, Stanley had beaten 195 self-driving competitors to win the $2 million grand prize. The fifty-person Stanford team was led by a professor named Sebastian Thrun, a German computer scientist whose research focused on robotic mapping.

Larry wore a white Google lab coat. As he dismounted from Stanley's bumper, he cast an aura of a mad technologist on the loose in his laboratory. He explained how Stanley worked. The vehicle used five LiDAR (light detection and ranging) sensors mounted on its roof to create 3-D maps of its surroundings, supplementing the GPS location, and fed this data into its internal guidance systems to control the car's speed and direction. Stanley banked on reliable 3-D maps to make its way across the expansive dry desert. Not surprisingly, good maps that include 3-D structures are a fundamental building block for self-driving cars.

A few months later, during the spring after CES, Google bought a Boulder, Colorado–based company called @Last Software in an effort to create 3-D cityscapes for the whole planet, something that would be necessary for the advancement and success of self-driving cars. Its software, called SketchUp, would be used to draw the photorealistic 3-D models you see today in Google Maps and Google Earth.

Keyhole knew SketchUp well. As with Keyhole, In-Q-Tel had invested in @Last Software. Its simple-to-use visual tool allowed users to quickly draw 3-D models of buildings and cities. SketchUp had done for CAD (computer-aided design) drawing what Keyhole had done for GIS: placed formerly difficult-to-use tools into the hands of the masses. The software's chief market had been architects, but because the tool was so easy to use, it had found favor in a variety of unexpected industries: stage design for the film and theater industries, woodworkers, and the DIY-maker market, to name a few. As with Keyhole, U.S. soldiers began using SketchUp to create 3-D models of hostile territories that they might get dropped into, which explained the interest from In-Q-Tel.

In early 2006, with the need to find photorealistic 3-D models of the entire planet for Google Earth, SketchUp's name surfaced again. John and Brian began to consider the prospect of crowdsourcing the effort: essentially getting the world to draw the world.

The CEO of @Last Software, Brad Stein, came to Mountain View for a series of meetings with the Google Earth team. These culminated with a meeting in Building 43 with Larry and Sergey. I remember the darkened room as Brad ran through a demo of SketchUp for Larry first, and then again for Sergey. John was hoping to create a partnership, supposedly, but the fact that he had invited Larry and Sergey to the meeting should have been my clue that he was hoping for something more. At the end of Brad's first demo, Larry said, "Maybe we should buy your company?" Brad nervously laughed off what he thought was a joke.

Stein's demo ended up being a test of the ease and usability of

SketchUp. It was about a week before Valentine's Day and Sergey asked Stein to help him on a project he had been struggling with. He was trying to make his girlfriend a gift. He wanted to 3-D print two interconnected hearts but coudn't find a software tool easy enough for him to draw the model. Brad used SketchUp to draw the interlocking hearts in five minutes, and emailed Sergey a file that could be used with a 3-D printer. I still laugh at the thought of the likely reaction of the girlfriend when she opened the Valentine's Day gift from her billionaire boyfriend. "Oh, Sergey, you shouldn't have."

The demo made the sale, however, and weeks later John prepared a presentation for Larry, Sergey, and Eric's staff meeting to ask permission to buy @Last Software. As John launched into his presentation, Larry interrupted, cutting John off. "I thought we told you to buy this company?" Larry said. "I don't understand why you are here."

"Well, it's a lot of money. I thought I should come ask," John said.

"Just go buy this company," Larry said.

John returned to Building 41 and said to me, "Well, that was easy." The @Last Software acquisition closed in March of 2006, and the team initiated Google's efforts to map the world in 3-D. (Today the models are extracted from sensors mounted on Google planes.)

I've returned to CES frequently since 2006, and the section of the show dedicated to the autonomous vehicles has grown exponentially over the years. Page's keynote, with Stanley's help, was arguably the starting gun for the technology industry's race to build self-driving cars.

That morning, Larry's second product announcement centered on a joint project with VW-Audi to embed Google search capabilities in a car's dashboard via Google Maps and Earth. The project was the first contemplation of what this user experience might look like: The full-scale prototype dashboard (minus the car) was wheeled out onto the stage. This exact same dashboard had been sitting outside my office—and John's—for the past six weeks, so he wouldn't forget about this time-sensitive project in development. Engineers from Google and VW had been frequent visitors to this dashboard as they scrambled to complete a prototype of an in-car navigation system—based on Google Earth—in time for CES. Brian McClendon and Michael Jones had flown into Las Vegas the day before Larry's presentation to work with him, rehearsing this part of his speech, and to oversee the completion and testing of the final demo code with VW. As Larry toured 3-D cities around the country and the globe, the hushed crowd was spellbound.

VW and Audi were not alone in working to bring Google technology to the automobile. During recent months, no fewer than eight car manufacturers—such as Ford, General Motors, DaimlerChrysler, and Toyota—visited Building 41, proposing similar in-car navigation projects. Almost all these companies had opened navigation development offices in Palo Alto, tapping the world's best software engineers for this important and profitable work. VW had moved the fastest, and the manufacturer had been chosen by John to create their prototype that Larry shared in Las Vegas.

Because there was so much interest from auto manufacturers and personal navigation devices manufacturers, a Google business development executive named Karen Roter Davis was assigned to negotiate deals with these companies. One day she invited me, John, and Brian to a meeting with Garmin, the Kansas-based leader of the personal navigation device manufacturers. Brian, a Kansan, was particularly keen to figure out a way of working with them. The Garmin reps brought a special prototype to Building 41 for us to experiment with. It was an external GPS device, a blue-and-gray plastic box about the size of a deck of cards, with no screen. (I still have it.) Instead of viewing your location on that Garmin device, you paired it via Bluetooth with your phone and viewed the location of the device on your phone. (In my case, a BlackBerry.)

This is hard to imagine (even for me), but that day during the meeting with Garmin, I saw something on my phone that I had never seen before: my location. I was a blue dot. Of course, I had viewed this plenty of times on personal navigation devices, like Garmin's popular Nuvi, but never before on *my phone*. Prior to this, if you wanted to see a map on your phone of your exact location, you needed to type in the address of where you were. There was no blue dot that automatically told apps your location.

Now, thanks to this external little brick from Garmin, I didn't need to tell my phone that I was in Building 41 on the Google campus. My phone knew where I was—even what part of the building I was in. I don't think I fully understood the implications that day—that this pulsing blue dot would soon launch an entire industry. To me, it was a novel and complex demo that required a panoply of independent technologies: an external device connected to a power source, paired via Bluetooth, and activated in a special version of Google Maps that I had to install on my BlackBerry device. It was a twenty-minute process just to get it up and running, if you knew what you were doing.

Ironically, my BlackBerry of 2006 did already have a GPS chip capable of doing what the Garmin external brick did, but BlackBerry did not allow applications to turn it on. It was just too much of a power drain on the battery to activate the dedicated GPS chip that was in the phone. My phone's GPS chip was intended to be turned on in emergency situations only. The inclusion of GPS chips into phones was a recent development. In the early 1990s, as mobile phones were widely adopted, 911 call centers began to be flooded with phone calls from people who had no idea where they were: trapped motorists, injured hikers, lost bikers, and a variety of other distress calls. Suddenly callers—many in great peril—could not tell dispatchers their locations. Per a 1996 FCC mandate, wireless carriers were legally required to be able to determine a 911 caller's latitude and longitude, and to transmit their position starting in 2000. (Note: During Hurricane Katrina, callers were forwarded from 911 dispatchers to the Coast Guard, but their location information was not forwarded, forcing the Coast Guard to devise their Google Earth–based workaround.) In order to comply with this law, in 2000, mobile phone manufacturers began shipping devices with GPS chips in them. As of 2006, no device manufacturer had built a phone that allowed the GPS chip in it to be turned on by third-party applications, like Google Local, for navigation or location searching. This blended GPS-plus-phone experience was probably what Garmin had in mind in early 2006 when they visited the Google Geo team. Of course, though they didn't know it and we didn't know it, there was a team down in Cupertino exploring similar ideas.

Garmin built this external GPS receiver in order to enable a seamless user experience while not draining the battery. There were significant hurdles in the way of this approach gaining mass adoption. But once you did get it running, the blue dot was indistinguishable from magic.

The blue dot followed you wherever you went. In April of 2006, I brought the Garmin prototype device with me on a trip to Texas and used it to navigate the 250 miles from Austin to my sister's beach house in Port Aransas. Shelley drove while I played with my new Garmin toy in the passenger seat; our two girls were strapped in the back. Watching the blue dot automatically follow me was intoxicating, especially when the map was in satellite mode.

It was an extraordinary user experience, though it was possible only with some digital duct tape to make the technologies work together. The separate Garmin GPS device sat on my dashboard, plugged into my cigarette lighter and then paired with my phone via Bluetooth, running the special build of Google Maps on my BlackBerry. "If this could all be shoehorned together one day," I said to Shelley, "it would be awesome to have this in your pocket." I watched in wonder as the blue dot sped along the wide open highways of the Texas coastal plains, seeing ahead to the grassy marshes and sandy beaches around the next bend.

Writing this today, in 2018, I find it hard to imagine a world without the ever-present blue dot. The truth is that a lot of work goes into representing where you are in all situations (including indoors where GPS does not work). Today the blue dot is still an area of research and imagination: Teams of engineers work to determine your real-time position with hyperprecision in a multitude of environments, down to centimeter-level accuracy, all the while using up the minimum amount of battery power. Now there is a group at Apple that is known internally as the Blue Dot Team. A similar group of engineers exists at Google, focused on enabling low-power ways for all sorts of third-party apps and returning information and services tailored to the blue dot. When you're not moving, it turns off GPS sensors and then wakes up when your accelerometer detects movement and supplements location by mapping you according to what Wi-Fi networks you are in range of and the relative signal strength of nearby cell phone towers.

For the third demo of Google Maps and Earth at CES, Larry spoke directly to the large audience. "I know many of you probably have BlackBerrys," he said. "Raise your hands." Over half the audience raised their BlackBerry devices into the air. "This is all something you can run on your phone now . . . you can get directions . . . I used this recently. I was lost someplace and I just used this. It was very helpful. You can see satellite maps. You can zoom in and out like you can on maps.google.com. I think these kinds of mobile devices are really starting to work, and I think that's an amazing thing."

Then Larry announced a version of Google Maps for the ubiquitous BlackBerry. Larry's device was projected onto the multiple screens throughout the convention hall, demonstrating the new Google Local for Mobile. He could move the BlackBerry's tiny screen by clicking the cursor arrows on his keyboard and toggling over to a satellite map, which would slowly load onto the screen.

Google Local for Mobile, however, didn't include the critical blue dot, which meant finding directions or businesses was much harder. If you have ever looked at a map in a mall or a museum, chances are you start with the "you are here" dot. It's the same way on mobile devices. Without the blue dot, you couldn't type in "Starbucks" without also inputting a location, like "Starbucks near intersection of Fifth and Lamar." To do that on a mobile device, with its tiny screen and even tinier buttons, you'd likely find it easier to keep your phone in your pocket and walk four blocks in any direction to find a Starbucks.

By 2006, the ingredients had been prepared: the blue dot; the mobile mapping application; the connected devices with the GPS chips. But no one had yet brought all these together into one elegant experience. No one had baked the cake.

As I stood in the Google booth watching the video feed, I remember being pleased with how much Larry talked about mapping-related technology. "This is good for our job security," I told John when I called him on the phone after Larry had concluded his presentation.

After my return from CES, I sensed other Google marketing managers picking up this geocentric vibe from Larry, too. Soon, like the eager CES attendees, Google marketing teammates asked me about potential roles on the Google Geo team. Their reaction to the keynote talk was similar to mine: Larry seemed to be expanding Google's mission to include organizing the world's geographic information, and Google Maps and Google Earth would provide users with the universal accessibility.

Looking back on it today, I can see how those 2006 CES keynote attendees caught a glimpse of the future that Friday morning in Las Vegas. The self-driving car industry may have started that very day: Within the year, Page hired Sebastian Thrun and many members of the Stanley team to lead Google's self-driving car project. Navigation would get smarter, more current, and more personal with Google Maps and Google Earth directions and search results. And while the iPhone was still a year away, Larry showed us connected smart devices with the very first mobile Google Maps on them. And the world would be centered around a simple blue dot, otherwise known as you. A world, it seemed, where we would never be lost again.

Chapter 16

OKAY, GOOGLE, WHERE AM I?

"Mr. Kilday, thank you for your generous offer," said the Realtor in a black dress suit, "but we are going to let you and Mrs. Kilday go home, as we have two other parties who have offered more."

I looked over at the real estate agent's glass-walled conference room; I had seen this movie before. Two other couples were waiting outside for their chance to make one final bid on the property. It was a face-off to see who would be willing to mortgage their future the most. Despite bidding $50,000 over the asking price for a 1,425-square-foot, three-bedroom home in Menlo Park, Shelley and I had not even made the final cut.

By 2006, this type of real estate bidding war was all too common in Silicon Valley. Drag the kids to open houses on the weekends, go to a Realtor's office on Monday night where sealed bids were opened up, and return home disappointed. Repeat. For the above property, which was listed at $1.5 million, there were eight bidders.

Something wasn't quite right: A family friend mortgage broker had preapproved Shelley and me for an amount that would have committed us to spend 66 percent of our combined take-home pay on housing. "I thought the rule of thumb was to spend under 35 or 40 percent of your income on housing?" I asked my friend. "Oh no, that line of thinking went out the window years ago," he said with a dismissive wave of his hand. People in Silicon Valley were well paid, but apparently they were willing to spend most of it on a house.

Our current 1,056-square-foot home was located in the redwood-tree-lined Willows neighborhood. It had worked for Shelley, Isabel, and me, but now Camille's area had taken over what had been a tiny office nook and things were becoming very cramped. Every new toy that arrived in the house required that one be given away. We could no longer buy six-packs of paper towels, opting for the two-packs that could be stowed under the kitchen sink. Getting dressed in the morning required visits to three different closets. I began to look at our family dog, Penny, in a slightly different light, as she took up eight cubic feet in the house. Could we possibly get a smaller dog?

Admittedly, a bigger house in Menlo Park would have been a stretch. My role at Google just wasn't at the "big house in Menlo Park" level. To me, there appeared to be a pecking order for housing up and down Highway 101. Individual contributors lived in San Jose and Santa Clara. Managers in Sunnyvale and Mountain View. Directors in Palo Alto and Menlo Park. Vice presidents and CEOs in Atherton and Woodside.

My marketing manager responsibilities not only came up short on the financial front, but also started to come up short as the day-to-day work became less intrinsically rewarding and fulfilling. As the Geo team grew and John added more horsepower to his staff (upward of hundreds of hires within months), my role became increasingly marginalized. Before, I had been the sole product manager on Google Earth. John now had fourteen product managers working on his team. One of my twenty responsibilities had been to manage the palette of icons in Google Earth; now one person was dedicated to just this job.

Google Local and Google Earth continued to expand—both in terms of coverage and countries launched and popularity. By March, we had launched Google Local in over twelve countries. I was left on the sidelines as John flew around the world, accepting innovation awards and buying up mapping companies. Back in Mountain View, I was also being left out of meetings with Larry, Sergey, and Eric (or LSE meetings, as they were called). John sheepishly told me, "I can't invite you, because we've been asked to keep the number of people in the room down." For each meeting, he could include one of his fourteen product managers, whichever one was the expert on the subject matter for the meeting agenda item in question. For example, he would bring product manager Kei Kawai in if the topic of discussion was mapping in Asia. Before all of these specialty experts, the core team of Keyhole had handled these issues as they happened.

It's only natural that the organization of an acquired start-up changes once it is inside a larger corporation. I wasn't the only former Keyholer whose role was being redefined. Dede was now a part of the Executive Admin team at Google, which meant that she was providing administrative support for executives other than just John. Lenette and her team were getting pulled into Sheryl Sandberg's Google Consumer Operations group. Noah, who held many roles at Keyhole, was also forced to pick one role and was subsequently placed on the Google Enterprise sales team. In contrast, other Keyholers' positions expanded rapidly.

Daniel had dozens of people all over the globe reporting to him, and eventually he reported to Megan Smith in Corporate Development. Michael became Eric Schmidt's favorite wingman and was being whisked around the globe on Eric's Gulfstream V to various conferences.

The inner circle that John came to rely on soon expanded to include dozens of Googlers, not just his Keyhole staff. The email group alias of the core team running Google Maps and Google Earth transitioned from kh-staff@google.com to geo-staff@google.com. The Keyhole email alias was retired. It was no longer relevant.

I gave up my product management responsibilities and redoubled my focus on a variety of marketing promotions. It wasn't viewed as the most important work within the corridors of the Googleplex, but I found the projects creatively satisfying even though the work wasn't the most strategic. That spring, I concentrated on two projects that would become the final punctuation to my time in Building 41.

The first project revolved around naming. As we rolled Google Local out in various countries, many of the regional marketing managers and country managers pushed back on the Google Local name. The direct translation of Google Local often did not work. "What is a Google Local?" asked my Spanish counterpart Bernardo Hernández before the launch of Google Maps in Spain. "I don't think we have this word with this meaning in Spanish. It sounds like Google Loco." I thought the name was crazy in English, too.

I continued to be a squeaky wheel about the Google Local brand. Since John was now the director of Google Geo, I thought we could make another run for a name change. Though he agreed with me, John was not very excited about reopening this contentious issue with Marissa. He had been awarded custody rights, but didn't want to rush to change the kid's name.

Through conversations with Debbie Jaffe, I got the sense that there might be an opening to consider switching the name to Google Maps. During the marketing planning for 2006, I sold the concept of hiring a San Francisco–based ad agency to help us create a campaign for Google's Geo products. I told Debbie: "We are going to spend millions of dollars building a brand here, and I want to be sure that we don't build a brand around the wrong name. Would Marissa let us do some testing of names to make a final and data-informed decision?"

John replied all to my email to Debbie and Marissa (just as I requested), arguing that "we should at least let Bill test the different naming options to see what the data shows." Marissa begrudgingly agreed, but listed out the naming options for this testing: Google Local, Google Maps, Google Maps and Satellite. This concerned me; I was almost certain that the two Google Maps options would split the ticket. Google Local might win out in a three-way race where its opponents were both using the preferred Google Maps name. Even though I hated the option, I was successful in getting Marissa to add a fourth option: Google Local and Satellite.

With the help of the product management team, we launched a test of the names in early March. For one week, 1 percent of visitors to Google.com saw one of four different links at the top of the home page. The results were clear: Both Google Maps options fared three times better than their Google Local options. The data was undeniable.

Yet Marissa still pushed back on changing the name. Tens of

thousands of advertisers were now a part of the Google AdWords for Local advertising product. They had signed up for their ads to appear as part of Google Local; as a result, this name change could introduce confusion for the advertisers. Worse, it might violate Google's contractual obligations to switch those ads over to a new product. Google Maps wasn't what they signed up for. Marissa said that a meeting would need to be called with all of the stakeholders to make this final naming decision.

Trying to get five or six high-level Google executives to commit to a meeting time was challenging. I used the reason of a "pending ad campaign" as a way to gather these hard-to-pin-down individuals into one room. I needed a decision by the end of March. A time and date were set. The meeting would be held on the afternoon of Friday, March 24, in Building 40 after a company-wide TGIF. The attendees included the CEO Eric Schmidt, the VP of Product Jonathan Rosenberg, the VP of Ads Jeff Huber, Marissa Mayer, John Hanke, and Larry and Sergey. Unfortunately, John couldn't make the meeting. He was scheduled to speak at a conference in Paris. We both agreed that the meeting should happen as scheduled since it was so difficult to schedule it in the first place. (I was not invited to the meeting; it was above my pay grade.)

I walked by the conference room to ensure that it was in fact going on. Peering through the door's window, I could see that everyone had shown up. Marissa sat at the head of the table and was speaking to the group as I walked by. I ran into Patti Martin, Eric's executive assistant whom I had worked with to coordinate the meeting's logistics. She said that the meeting had gotten a late start, but that they had been in there for over an hour. Patti knew the subject of the meeting and how important the outcome was for me. She said that I looked like an expectant father in a hospital waiting room, anxious to hear if it was a boy or a girl. She told me to go on home (it was nearing six o'clock) and that she would call me with the news once the meeting concluded.

By the time I returned home to Menlo Park, I received an email from Patti. The meeting had let out around 6:15 P.M., but they left the conference room tight-lipped. No one told her the name before leaving for the weekend.

Immediately I emailed Debbie, who was also awaiting the news. She said that she would try to get ahold of Marissa. I emailed Debbie again at around 9:30 P.M. and asked if she had heard from Marissa about the final decision. She said that she had heard the news, but couldn't share it with me. She did tell me that Marissa wanted to know when John was scheduled to return to California. I dialed Debbie's cell phone number. She picked up and said, "Look, I know you want to know the name, but Marissa made me swear that I wouldn't tell anyone. She wants to deliver the news directly to John. You know that she and John haven't had the best of working relationships, and she wants to use this as an opportunity to reach out directly to John."

This was all true.

"Oh, come on, Debbie. I've worked hard on this. You have got to tell me what the name is. I promise that I won't tell John."

"No, I can't. Marissa wanted me to find out from you when John gets back, and she explicitly told me not to tell you." I told her when John was scheduled to return, and I begged her to tell me once again. Finally she relented.

"The name is Google Maps," she said, "but you *cannot* tell

Hanke, okay? Marissa is going to call him this weekend with the news." This was the reason why the meeting had dispersed without a word. If this sounds to you like I'm describing a junior high who-likes-whom rumor mill, I can assure you that it gets worse.

Earlier in the evening I had received an email from John, who was up late in Paris. "Email me if you hear anything," he had said. After speaking with Debbie, I emailed John to let him know that he was supposed to be receiving a call from Marissa this weekend.

The next morning around eight-thirty my phone rang. It was John, from Paris. I hadn't had my first cup of coffee yet.

"Did they decide the name?" he asked.

"They did," I said. "Apparently, Marissa is going to give you a call to tell you what the name is. I spoke with Debbie last night because they wanted to know when you got back to California. Marissa is going to call you tomorrow."

"Does Debbie know the name?"

"Uhmm . . . well . . . yeah . . . I think she does," I replied.

"Did she tell *you* the name?"

"I'm forbidden from telling you. Marissa is going to call you directly to tell you the name herself. She wants to tell you herself, and I had to swear up and down to Debbie that I wouldn't tell you the name," I said. I was now fully awake.

"Bill, what's the name?"

"John, I can't. Marissa is calling you tomorrow."

"Bill, just tell me the name." His tone changed. It wasn't a friend talking about who liked who in the eighth grade. This was John, the director of Google Geo, asking—no, demanding—to know the name of one of his two central products.

"Okay, I'll tell you, but you have to promise that you will act surprised when Marissa calls you tomorrow."

"Okay, I'll act surprised," he said a bit unconvincingly.

"No, John, you really have to act surprised, okay?"

"Okay, okay, I'll act surprised," he said impatiently.

"Google Maps."

"All right. Thanks. I gotta get a plane now."

My phone rang again at around nine o'clock on Sunday evening.

"Why did you tell him!" Debbie started in on me. "I told you not to tell him! And you told him! Marissa called me and was really pissed!"

I explained to Debbie that John had called and asked me what the name was. I told him that I couldn't tell him, but he demanded. And John had promised to act surprised when Marissa called.

"Well, he didn't!" Debbie recounted for me what Marissa told her. She had called John per the plan and told him that she wanted to share some good news.

"The decision was made to go with Google Maps as the name," Marissa had related.

John replied with feigned surprise, but apparently Marissa didn't buy it. "Bill Kilday already told you, didn't he?"

John didn't respond quickly enough and had to admit, "Yeah, he did."

We pulled the trigger on the public name change almost immediately, as we didn't want to leave room for anyone to change their minds. Elizabeth Harmon pushed the change on April 2. Google Maps was now Google Maps again. There was no more Google Local.

Getting the name changed to Google Maps had been a hard-fought victory and I wanted to make sure the name could never change again. I wanted it to be *permanent.* I wanted it to be carved in stone, dammit.

Just for fun, I posted a random ad on Craigslist looking for a sculptor to carve a product name into stone. In a few weeks, I had the Google Maps logo, completed in Catull (the font of the Google logo), carved into a beautiful eighty-pound block of pink alabaster by a Berkeley-based sculptress. Noah picked it up for me, since he lived in the East Bay, and delivered it to a pedestal in Building 41. It's a beautiful piece of art and still sits in the Google Geo offices today. Yes, it cost the company $1,300; but it represented a final decision. If anyone *ever* tried to change the name, my thinking went, we would simply say, "We're sorry, we can't change it. The Google Maps name is carved in stone," and simply point to our beautiful eighty-pound hunk of pink alabaster.

In what ended up being my final project completed in Building 41, I planned an event for developers using the Google Maps API. Other than small meetups and hackathons, Google had never done a large-scale, developer-focused event. My idea was to host an event on the Google campus for engineers, web developers, and designers using the Google Maps API. I named the event Google Geo Developer Day, and Jens Rasmussen helped with the map branding and design elements for the event.

While it was nearly impossible to get money for marketing outlays or advertising, the idea of an event that would target developers found quick favor with Google executives, and the budget and concept was approved. At an engineering-oriented company, spending money on engineers was an easy sell. Bret Taylor was very enthusiastic about the idea.

Held in June of 2006 on the Google campus, that first developer relations event attracted around three hundred Google mapping enthusiasts, developers, and press. We framed the event as a sort of launching pad for several key mapping initiatives, including photorealistic textured 3-D buildings in Google Earth (courtesy of the contributions from Google SketchUp power users). Bret and Jim Norris showcased a new Google Maps API for Enterprise product, which allowed companies like Zillow, OpenTable, and Yelp to pay to use Google Maps on their websites in exchange for more favorable terms of service, including a promise that we wouldn't start showing ads for competitors on their sites.

I invited Larry, Sergey, and Eric to the event. They were surprised (and I was surprised that they all showed up) by the attendance, enthusiasm, press attention, and fun at the event. During the conference lunch hour, I instructed all of the attendees to lie down on the grassy lawn outside of Charlie's Café. I had arranged for a low-altitude aircraft to fly over Google and take high-resolution aerial images of all the attendees out on the grass. It was our personal *Powers of Ten* moment. The photos were downloaded, printed, and distributed to the attendees as a memento at the end of the day.

The event was a hit, and widely recognized at Google as a model platform to launch new products and product features. This was a first for this sort of a product orchestration, with the marketing team dictating the specific date of a launch versus the products being rolled out simply when they were ready. The following year, the event was broadened to include more than just Google Geo products, and it was called Google Developer Day. In 2008, the concept was expanded further to include its own unique brand, Google I/O, which is the company's largest annual developer relations and product launch event. Held annually at the Moscone Center in San Francisco, attendance to Google I/O is capped at five thousand, and tickets sell out within hours of going on sale. Today teams of people at Google work on this event.

After the success of Google Geo Developer Day, however, I was spent. Burnt out. I had left it all on the field. Shelley was getting tired of the pace, too. We had recently gone to Disneyland with our daughters, and I had spent the entire day on my BlackBerry.

I was ready to return to Texas. I can't say it was any one thing; it was likely a combination of factors. It had been an intense six-year ride. My role was now purely a marginalized marketing role, and Google was pretty anti-marketing at the time. It was all about the product. Not that I could blame them; the strategy of simply building dynamic user experiences had worked—and continues to work. Shelley and I questioned if Silicon Valley was the right place for us to raise our girls. I wanted to raise my kids around some below-average kids, too: the ones watching DVDs on the five-minute drive to school while they eat chocolate donuts in the back seat. Though we would miss our extended family in California, the girls would get a chance to spend more time with my large, zany family in Texas.

While I was considering the possibility of leaving Google and Mountain View behind, as luck would have it, there was a chance to return to Austin—*and* still work for Google. In the spring of 2006, I was frequently pulled into meetings with large companies spending millions to advertise on Google. There was always a parade of marketing managers like me in the room, but my Google Maps and Google Earth demos were often saved for last. There was no better example at Google of the rapid pace of innovation. One subset of companies that lined up for meetings at Google were PC manufacturers: Hewlett-Packard, Dell, Acer, Sony, Toshiba, and others.

During these meetings, I frequently crossed paths with a Googler named Sundar Pichai. He was a product manager for two products: Google Toolbar and Google Desktop. He reported to Marissa; at this point in his career, he was a product manager, not even a director. Sundar was extremely friendly and, even by Google standards, extremely bright. He was also uniquely customer-service-oriented, which was rare at Google. I think that came from the fact that he had worked at McKinsey & Company as a consultant, so he knew what it meant to have an actual customer whom you have to please.

All of this experience made Sundar effective at creating distribution partnerships with PC manufacturers, including one that appeared in my Google newsfeed one Friday night with Austin-based Dell Computer. Google was investing heavily in distribution partnerships with companies like Dell, Apple, HP, Mozilla, and others, to combat the threat of Microsoft automatically switching the default search provider on hundreds of millions of devices over to its own search engine, Bing. With 85 percent of all web browsing being done on Microsoft's Internet Explorer, this threat was very real. Products, like Google Toolbar, running as a helper app to Internet Explorer, allowed Google to guard against Microsoft's ability to automatically switch hundreds of millions of users away from Google.

I immediately emailed Sundar and told him that if he needed anyone to "work on the partnership," I would be happy to move to Austin to help. Though the deal size was never publicly confirmed by either company, CNN reported that Google paid Dell $1 billion to be the worldwide preferred search provider on the 47 million PCs that Dell shipped annually. It was a three-year deal, and I was curious to know if Sundar might need a point person in Austin to ensure the partnership went smoothly. He replied that he hadn't thought that we needed someone in Austin, but that he would consider it, and forwarded my email on to the hiring manager with a note recommending me based on my Google Maps and Google Earth experience. Within two weeks, I had a job offer, complete with a relocation package to Austin. Shelley and I began to look at houses and schools there.

I talked to John about the opportunity, and he answered as a true friend would, not as my boss. "You should think about all of the people you are working with here at Google," he said. "They are going to be sprinkled all over Silicon Valley one day as CEOs and chief marketing officers. So you'd be leaving that whole professional network behind." He also acknowledged that the new role might be more strategic to Google than my current Google Maps and Earth marketing role. And having lived in Austin himself, he had to admit, "But if you can figure out how to live in Austin *and* still work for Google, that's pretty hard to compete with."

In October 2006, I took the job in Austin. I'm not one for long goodbyes; I turned over the marketing team to Jeff Martin, a very capable head of marketing from @Last Software, and hit the road. Dede organized a surprise party for me; it was such a surprise that I wasn't even there. I was on a flight bound for Austin. When the plane's wheels hit the tarmac, there were eight voice mails from Dede and others asking where I was. The team threw a going-away party without me, sending the pictures of the globe-shaped cake that Dede had made. My time working for John was over (or so I thought). I was sad to move on from the team, but immersed myself in my new Google role in Texas.

Chapter 17

STREET CRED

During the coming months, I settled into my new Google position, working by myself out of my 3,800-square-foot house in the Pemberton Heights neighborhood (we bought it for the same amount that we sold our 1,000-square-foot house in Menlo Park). One afternoon, I called HR at Google with an inquiry.

"Can you give me a list of everyone who lives in Austin, Texas, and works for Google?" And the HR representative responded, "We don't have anyone in Austin as their office address." And then I asked, "Could you see if anyone has a 512 area code?" She came back with three names. I emailed these individuals—no one knew the others even existed—and organized monthly barbecue lunches for the group. Soon we shared an office in downtown Austin, right next to Waterloo Records on North Lamar, quickly growing from four to six to ten Googlers and then the establishment of a Google Austin office. My Texas take on the Google mission was emblazoned across our office T-shirts: "Organizing the world's information, y'all." (Now the Austin office is home to about 750 Googlers.)

The partnerships job wasn't the best fit for a marketing guy like me; I spent more time looking at spreadsheets and contracts than I wanted to. To be honest, I wasn't particularly good at the job. My heart and passion were still in marketing. That said, I ended up working on some of the most strategic partnerships for the company, including Apple, Mozilla, Dell, and Adobe, among others.

The Google search box wasn't just available on the home page of Google.com, it was embedded in dozens of other products, like Apple's Safari, Mozilla's Firefox, and Internet Explorer, and I went to work on these partnerships. Literally billions of dollars flowed through these little distributed Google search boxes. Firefox alone—whose default home screen featured Google—generated billions in search revenue annually. If there had been a list of Google employees ranked by revenue, I certainly would have been near the top. When John updated me on some new outlandish mapping project (such as purchasing a mapping company in Italy or an exclusive data deal with a company in Asia), I was quick to point out that I was now funding his Google Geo division expenditures. While Larry and the rest of the executive team pressed John and Brian to expand and grow in every way imaginable, they still did not prioritize map monetization: Google Maps and Earth continued to spend much more money than they earned.

It was satisfying to be on the other side of the counter again—actually generating revenue instead of spending Google's money with reckless abandon. If I thought the rate of investment in Google's Geo team was lavish in 2005, the spending swelled to unprecedented levels by late 2006. The budgets for John's area appeared to be boundless. @Last Software was purchased for a rumored (though not confirmed) $45 million in February. Google locked up tens of millions of dollars' worth of Digital Globe satellite imagery. It bought another mapping company called Endoxen, based in Lucerne, Switzerland, in order to expand the company's European footprint. It was clear that Google Maps and Google Earth offered superior user experiences. Wherever we could get the data to launch in a country, Google products would immediately garner significant market share. By the end of 2006, Google Maps had launched in forty-seven countries, and Google Earth had been downloaded over 120 million times.

It's only natural to think that when you leave a longtime job you will be missed and it will be hard to fill your shoes. *Certainly the team will crater in my absence.* I entertained no such illusions about the Google Maps team after my departure to Austin. I knew that John and the team would be just fine without me. Truthfully, the path of Google Maps was still being paved.

Luc Vincent joined Google in 2004. Because of his background in "computer vision" (computers with an ability to see), he was assigned to Dan Clancy's team, with the monumental task of scanning the contents of libraries throughout the world. It was a remarkably ambitious project and assembled dozens of the industry's leading computer vision engineers. Internally, it was code-named Project Ocean.

Soon after he joined the team, Vincent was summoned to a meeting with Larry Page. Page told Vincent about an entirely different project, one that he had started personally with a computer science professor at Stanford named Marc Levoy. The project involved capturing videos of cities from street level, then creating a continuous strip of horizontal images and extracting data from those images (for example, an address) to make them searchable. Vincent understood swiftly that Larry was enamored with the long-range concept of making the *physical* world searchable, not just the digital world of web pages. Google had made a grant to Levoy to fund the development of a proof of concept in 2003, and Levoy asked Larry to renew the grant in 2004 in order to continue the project.

To manage this relationship, Larry asked Vincent if he would act as a liaison with Levoy and his students on the project to assess the progress and provide a bit of direction. At its core, Larry explained, Levoy's proof of concept relied on many of the same computer vision ideas being employed to scan books with page-turning robots; both projects involved taking pictures, stitching those pictures together, and then extracting searchable data out of those pictures.

As a part of this conversation, Vincent told me that he asked Larry how the data that Levoy was using had been captured; he expected that Google had found some supplier of the imagery. "I just drove around one Saturday capturing the video myself with my camcorder," Larry explained. Then he showed Vincent the video on his computer.

After canvassing the Stanford campus, the footage showed Larry and his two friends driving west to Half Moon Bay on the winding Highway 92 and then up Highway 1 to San Francisco. Viewers of this video, including Vincent and Brian McClendon, claim that Larry can be heard on the audio bantering with his friends, Marissa Mayer and Sergey Brin.

This just kills me. For a minute, imagine yourself standing on a corner in Palo Alto some random Saturday afternoon during the summer of 2002, and a car slowly rolls by with a guy hanging out of the passenger window with a camcorder steady in his hands. What would you have thought? Would a friend believe you if you had turned to her and said, "Hey, did you see that? That was Larry Page, Marissa Mayer, and Sergey Brin. They just drove past in that car right there and, weirdest thing, I swear Larry Page was taking a video of me as they drove by," would your friend have believed you?

The concept of a street-level perspective of a city photographed and seamlessly stitched together was not new. In 1979, a project funded a group of MIT researchers, including Nicholas Negroponte, to create the Aspen Movie Map. The team utilized four sixteen-millimeter cameras mounted on a car with gyroscopic stabilizers. The movie map snapped photos every ten seconds. The technology also provided an overlay map that allowed users to control the direction they traveled in the virtual tour of the city.

In 2004, Larry's movie had become the basis of Levoy's street-level mapping proof of concept. Vincent took on the liaison role with the team of Stanford computer science students in the fall of 2004 as his "20 percent project." Google engineers are allowed and encouraged to work on projects of interest to them on a 20 percent basis (Gmail had famously begun as a 20 percent project in 2002). So Vincent kept his role as a computer vision engineer on Google Books while also collaborating on the project with Levoy at Stanford.

By the spring, Vincent began to get a clearer idea of Larry's level of ambition for the project; Larry periodically stopped by Vincent's cube to check in on its progress. That summer Vincent recruited seven additional Google engineers, including an electrical engineering PhD from Stanford named Chris Uhlik, and seventeen interns (many from Levoy's class) to work with him. They built the first car prototype, captured their first images, and developed a "computer vision pipeline" application. Capturing the images was part of the problem; there was an equally daunting task of managing, mapping, and mosaicking all of these 360-degree images together. It was like a more sophisticated version of Mark Aubin and John Johnson's data processing tools.

The first Google Street View vehicle likely generated some alarm throughout the streets of Mountain View that summer. The unmarked dark green Chevy van with an assortment of mutated computers, cameras, and laser sensors strapped to its top couldn't go faster than ten miles per hour because the captured images would be blurry and rendered useless. After multiple blown fuses in the van, a separate Honda gasoline generator was bolted to the roof to provide more reliable power.

The van proved to be terrifically unreliable. Every day it was readied and its mission mapped out, and all of its systems carefully restarted and connected. And every day it would cruise for about an hour before the intern driver would call back to Mountain View to report some sort of computer crash or system failure, and he would have to return to the mother ship to diagnose what had gone wrong.

By the end of the summer of 2005, however, the Google Street View project was anything but a failure. The team successfully captured its first data set of the streets of Mountain View and Palo Alto, and even managed to integrate the imagery into Google Maps as a demo.

In October 2005, Vincent and Uhlik presented a tech talk in

Building 40. It was well attended by many on the Google Maps team, including John, Brian, and me. The Street View team still represented Vincent's 20 percent project, and since he was functionally reporting in to Dan Clancy and Google Book Search, Street View was considered under the umbrella of Google Books. This remained the case because, at least initially, Brian remained skeptical of the long-term prospects of Street View and was therefore happy to have that team report to someone else.

I shared Brian's skepticism of the whole scheme: I had done the math on what it was going to take, and it added up to a ridiculous amount of time, mileage, gas, energy, and fleets of cars and drivers to create a worldwide data set for Street View. According to my back-of-the-envelope calculations, it would cost hundreds of millions of dollars to develop. In addition, assuming one could capture all of this street-level imagery, I didn't see how this kind of data could be elegantly integrated into Google Maps.

Despite the skepticism, after the tech talk in Building 40 and a positive engineering review with the director of Google Engineering, Bill Coughran, Street View became a formal project at the company, a significant milestone for any initiative. This translated into budgets, recruitment, legal clearances, office space, server allocation, timelines, agreed-upon deliverables, competitive analysis, privacy reviews of data, and more. In addition, the Street View project moved from Google Book Search and into Google Geo, reporting up to Brian.

By the summer of 2006, Vincent had hired twelve full-time engineers and rebuilt the whole system—the cameras, the cars, the processing tools. That said, Street View was still considered by many at Google to be an experiment, though an experiment without limits. The team had no budget constraints and Larry invested heavily in the concept.

John introduced Google Street View to the world at the Where 2.0 mapping conference on San Jose on May 29, 2007. Even though the launch included only five cities (San Francisco, Vegas, Denver, Miami, and New York City), Vincent's experiment was an instant hit. The reception at the launch created a server bandwidth spike that came within single-digit percentages of topping even the most optimistic demand projections. Overnight, Brian went from being skeptical about the project to pressuring Luc's team to get more cities covered with Street View imagery as soon as possible.

However, scaling the concept to cover millions of miles, instead of thousands of miles, was a completely different challenge. Vincent and Uhlik's Street View proof of concept patchwork was too complex, hand-customized, and unreliable. It was never going to work on a large scale. Thankfully for John and Brian, an entirely new team appeared on the scene in early 2007.

Enter Stanley, stage left.

You remember Stanley, right? Since their 2005 DARPA Grand Challenge victory, Sebastian Thrun and his Stanford all-star team of robotics, computer vision, and self-driving car PhD candidates had been hard at work creating the next generation of autonomous vehicles. In 2007, Thrun proved equally creative at navigating complex business and legal arrangements. He started his own venture, independent of Stanford, and then offered to sell that new company to Google (noting that Microsoft was also interested).

Google bought Thrun's company for a tidy sum, and that team

was joined by another star in the self-driving-vehicle community named Anthony Levandowski, who was a master's student in the UC-Berkeley Industrial Engineering and Operations Research program. Thrun had met Levandowski at the DARPA event in 2005. Levandowski's self-driving motorcycle, Ghostrider, had failed to complete the rough desert course (Ghostrider is now displayed at the Smithsonian). Levandowski joined Thrun at Google in 2007, first working on Street View, and then going on to lead Google's self-driving car project.

In April of 2007, Thrun's team and Vincent and Uhlik's group restarted the Street View project from scratch. Together they designed the next, simpler generation of Street View cars, with a goal of mapping the United States as quickly as possible. (Thrun's team was given additional lucrative earn-outs tied to reaching an aggressive goal: collecting street-level imagery for one million miles of the six million miles of U.S. roads.) The new car design was much more basic than Vincent and Uhlik's prototype vehicles: It utilized off-the-shelf, high-end cameras and didn't require more complicated laser sensors or moving parts in the camera apparatus. Vincent and Uhlik's learnings and initial work helped create the first five cities, but it would be Thrun and Levandowki's elegant and simple design that would allow Google to scale the Street View coverage to what it is today.

A fleet of second-generation Street View Subarus fanned out across the country—from the congested avenues of Manhattan to the leafy suburban roads of small midwestern towns. By late 2007, one million of the six million total miles of U.S. roads had been driven and captured. Thrun's team met its aggressive goal.

The general public's reaction to Google Street View—and

the iconic orange Pegman in the user interface—was more than enthusiastic. And much of this frenzied response was captured by the passing cameras. Mad acrobatic feats were performed upon the sight of Google Street View cars in Florida. Pedestrians dressed up in scuba gear (complete with flippers) and chased the cars down the streets of Norway. And there were various scenes of people caught in the act of something they would prefer to not have the rest of the world know about. Due to privacy concerns, all captured images were run through a computer algorithm in order to search the images and blur cars' license plates and people's faces. Unfortunately, the Google blurring algorithm worked only on faces.

I called John to pitch him on a concept for an ad campaign for Google Street View. While I was no longer in charge of marketing for Google Maps, I was still grandfathered into being able to pick up the phone and pitch John an idea or two.

My idea was this: Pick a city where Google Street View cars were scheduled to map its roads in the near future. Then take out in advance a full-page newspaper ad in that city. For example, if the team was planning to capture Austin, the ad's headline might read AUSTIN. GOOGLE IS COMING. This announcement might be positioned above a color photo of the cute Google Street View car. Underneath the car, there would be a second headline: LOOK BUSY.

John didn't buy it. "Kilday, can you imagine the chaos that it would set off? The political causes? The crazies chasing the cars down the street wearing scuba gear? You'd have full city blocks lined up with flashers!" he said.

"Exactly!" I said.

I still wish he would have done it.

I remained skeptical of the Google Street View project. I often shook my head in disbelief when John or others shared with me the sheer scale of the operation. The drivers, the hundreds of cars, the millions of miles driven. As cool as the user experience was, I didn't understand the fundamental economics of it all. Then again, maybe I was asking the wrong question. Maybe it was wrong to apply traditional business logic to such a nontraditional company.

When Google had gone public in 2004, Larry had cautioned buyers of the stock in an open letter entitled "An Owner's Manual for Google Shareholders." "Google is not a conventional company. We do not intend to become one," he wrote. "Sergey and I founded Google because we believed we could provide a great service to the world—instantly delivering relevant information on any topic. Our goal is to develop services that improve the lives of as many people as possible—to do things that matter."

As the scale of Google's mapping operations under John's leadership continued to expand, it reinforced for me the true meaning of the owner's manual.

Chapter 18

4,000 LATTES

From: steve@apple.com
To: jhanke@google.com

> Hi John. I was wondering if you would be available to meet?
> Steve Jobs

"So, what do you think the chances are that this is real?" John asked his wife, Holly.

High in the hills of Piedmont, California, John and Holly were spending a lazy Sunday afternoon reading the newspaper and lounging around the house. It was a mild fall day in 2006. As he often did, John opened up his MacBook to check his email, and one message in particular caught his attention.

"About 10 percent," Holly said. Despite her assessment, John replied to the email, offering up a window of time between eleven o'clock and noon the next morning on the off chance that the email was sent by Steve Jobs.

At 11:05, John's phone rang in his office at Google.

"Hi John, this is Steve."

Jobs was complimentary of the work that the Google Geo team had done to date, and began to tell John a little bit about a new project for Apple. "You may have heard rumors about a device that we are working on. That device may or may not have mobile capabilities," Jobs guardedly explained. "We'd like to talk to you about working together. Would you be interested?"

As a teen in Cross Plains, John had idolized Steve Jobs and Apple's magical Macintosh computers, even if in 1984 he could not afford one. By late 2006, there was widespread speculation about Apple's as-yet-unnamed device. No one outside of the secretive halls of One Infinite Loop had yet seen the device, but the Apple rumor community had primed the technology world with high hopes. John too was primed by the prospect of being involved in this highly anticipated new mobile device and promised Jobs that he would personally manage the project.

The kickoff meeting was held on Tuesday, October 31, 2006, at Google's offices. John escorted the Apple software executives, including head of Apple software Scott Forstall, into a conference room for the start of this world-changing technology project. In the spirit of Halloween, Google's lead server engineer was dressed as a nun in a black robe and white habit. During this meeting, they discussed how Apple developers could tap into Google's back-end mapping services.

The front-end Google Maps app that was installed on the new device ended up being relatively simple to create, as Apple had clearly designed the device to make application development easy. Much of the heavy lifting was being done by Google mapping back-end services. All the street data, the driving directions, the local search results, the addressing, and the satellite imagery was served from Google to this new front-end mobile application (or app, as they called it) that was running on Apple's new device.

On January 9, 2007, when Steve Jobs stepped onto the stage at the Moscone Center in San Francisco for his seminal product introduction, it was the first time anyone at Google ever saw the iPhone. Not even Eric Schmidt, who was on the Apple board of directors, had seen "the device." As a witness to technological history, John sat in the front row for Jobs's monumental presentation: the introduction of the iPhone with its beautiful large multi-touch screen so perfectly suited to run Google Maps.

"I want to show you something truly remarkable," Jobs said, late in the demo. "Google Maps for iPhone."

Jobs clicked the Google Maps icon on his phone, and a blue dot pulsed with his current location. The app had access to the GPS data and automatically centered the map view over Moscone without Jobs needing to input his location. He then entered the first public search on the iPhone, typing in simply "Starbucks," bringing back a beautiful animation of fourteen map pins dropping onto the map. To showcase the revolutionary integrated calling capability, he clicked that trademark Google Maps map pin icon and called that Starbucks directly from the map, which no phone had ever done. Jobs ordered 4,000 lattes from the surprised barista before apologizing and hanging up to thunderous applause and laughter.

"Now let me show you something truly amazing," he continued, switching over to satellite mode. Jobs zoomed from space down to the Moscone Center on his iPhone, panning over to the pyramids of Egypt, then to the Eiffel Tower and to the Statue of Liberty. Jobs, and the entire Moscone audience, fell silent as he toured the globe with childlike wonderment. This audience was spellbound by the magical force of the technology. Was it possible that this was running on a phone? A tour that looked a lot like the one John had demonstrated for Shelley and me at our house in Austin eight years earlier. *The superman thing.*

"Isn't it beautiful," Jobs said.

And it was truly remarkable. It proved to be the killer app of the iPhone. It was a revolution in the way the world found its way in the world. Fast and fluid and visual: a Google Map that you could simply push and pull and pan and zoom.

In my new strategic partnerships job in Austin, among many partnerships, I was responsible for reporting to Apple on traffic and revenue, so I had access to the traffic by type of device (for example, iPhone versus a desktop versus iPad). I also knew from industry reports the number of iPhones that had been shipped, so the math made it clear to see the usage patterns. From the early days of Google Maps and Google Earth, I knew users of desktop or notebook computers might call up a map once or twice a week. But when I looked at the aggregated iPhone traffic, I saw Google Maps was being called up *once or twice a day.* (It's important to note: All data was aggregated and anonymous. I had no access to any personally identifiable information.) It seems logical to me now: Your iPhone is always with you throughout the day—and it is such a convenient way to find your next destination. The usage patterns were astounding to witness.

The first day you could actually buy an iPhone was six months after Jobs's demo: June 7, 2007. Within one month, in terms of Google Maps usage, the iPhone surpassed the entire installed base of all other Google Maps for Mobile builds. Within eighteen months, the usage of Google Maps for the iPhone surpassed that of Google Maps on desktop and notebook computers.

Let me repeat this: Within eighteen months of the iPhone's launch, more Google Maps usage was happening on that one device than on all other computers and all other phones *combined.* For years, PCs were being distributed throughout the world, numbering in the hundreds of millions of installed users—and the traffic from the iPhone passed them all by in a year and a half. And this was when the iPhone was available on only one wireless carrier network (AT&T) and available in only a handful of countries.

In many ways, it was the realization of the ridiculous slide in John's original pitch deck to potential Keyhole investors. Eight years earlier, I had scoffed at the idea as total Photoshop-ware back in San Diego: the idea that one day rich interactive maps would be accessed on pocket-size devices.

During the course of the demo, Jobs had also served Google and others a warning. A warning that would signal the start of a battle between Google and Apple over maps innovation for years to come. At the heart of the incredible user experience was the core innovation that enabled the user to quickly zoom in and out without the use of a keyboard; the feature was called multi-touch. It allowed the user to zoom in and out using two fingers to pinch and pull the map. Without a keyboard, this functionality was hugely important for the all-glass touch-screen phones. As Jobs called out and reinforced in his slides (a point that was probably made specifically for John, who was sitting in the front row), multi-touch was "highly patented."

Of course, Jobs and Apple were not the only ones developing a new generation of mobile device. In 2005, Larry had noted the proliferation of various phones and Google had bought a small mobile operating system company based in Boston called Android, headed by phone operating system wiz Andy Rubin.

In the spring of 2007, back in Austin, one of the three other Googlers I officed with was an engineer named Jeff Hamilton. Jeff was working on a secret project: He would say only that it was "software for phones." He was a part of Rubin's small Android team that Google had acquired in 2005. What Jeff didn't tell me was that the "software for phones" was actually a brand-new mobile operating system for smartphones.

By then, there were about twelve mobile operating systems, led by Symbian, Windows Mobile, Linux, and BlackBerry. Larry Page complained about the mobile team needing over a hundred mobile phones to test Google services. Google bought Android, two years before the launch of the iPhone, in the hopes of creating some order to the chaotic fragmentation. Namely, Larry wanted to open-source the software around a uniform platform and a set of APIs.

I saw Jeff testing a little white BlackBerry clone. Though he never told me what he was up to, I got the sense that he was working on some phone that was going to be branded as a Google phone. He even showed me a tiny Google Maps app running on its tiny little screen (60 percent of the face of the phone still had a QWERTY keyboard). After the launch of the iPhone, suddenly Jeff's prototype phone looked, to be honest, sorta sad. It was like Google Maps from a hundred years ago. Not surprisingly, the 2007 iPhone announcement made the project dead on arrival. Rumors circulated that two years of development by Rubin's Android team were completely scrapped. They were forced to start over from square one.

In order to define and lead a strategy for all mobile devices that were *not* Android, Larry hired an executive named Vic Gundotra away from Microsoft. A fifteen-year Microsoft veteran responsible for software developer relations, he was a smart, articulate, and savvy web services and software executive. John and the Keyhole team knew Gundotra from his early evangelizing of EarthViewer within Microsoft developer circles.

When it came to Google Maps for Mobile, there was clear overlap between John's world (maps) and Gundotra's newly defined world (mobile devices that weren't Android). A difficult question quickly emerged: Who should own Google Maps for Mobile? The product was a map, so should it be Hanke? But it was on a mobile device, so should it be Gundotra?

About a year after the iPhone announcement, Gundotra grabbed the reins of the Apple project. But Gundotra's time on Google Maps for the iPhone turned out to be short-lived. The contract with Apple was up for renewal, and, for whatever reason, Gundotra's approach had struck a negative chord with the wrong person in Cupertino.

By late November of 2007, there was still no renewal deal in place. With the risk of the whole project imploding, John was visited by Scott Forstall and Philip Schiller, heads of Apple's software and marketing respectively. Apple didn't want Gundotra on the project anymore and threatened to go their own way: to create their own mapping application. (This was a naive threat: Apple was already underestimating what it would take to create its own mapping service.) To reinforce their point, Forstall and Schiller requested that John meet with Steve Jobs himself.

"If fucking Vic Gundotra steps foot on this campus, I will

personally and physically remove him from the building," Jobs started the meeting. "Actually, I don't want him to come within a mile of this campus. And I refuse to even look at a contract that is more than one page long." Jobs had thrown in this last demand for good measure.

There was a bit of irony in Jobs's machismo. Being terminally ill with pancreatic cancer, Jobs weighed, by John's estimation, about ninety-five pounds. But there was no mistaking the visionary's passion and demands. Gundotra stepped off the project, and the deal got done. The contract was two pages (the deal between Dell and Google that I worked on was, by comparison, eighty-seven pages).

This proved to be but the first of many skirmishes between Google and Apple. As Google rolled out its Android-based phones, Jobs accused Rubin and Google of stealing many of the iPhone features. At an Apple developer relations event, he characterized Google's mantra of "Don't Be Evil" as "It's Bullshit." Apple sued Android device manufacturer HTC for implementing a multi-touch screen. Google CEO Eric Schmidt stepped down from the Apple board of directors, citing too many overlapping areas of operation between the two companies. And eventually Apple launched its own (disastrous) maps.

Indeed, a battle royale over ownership of the smartphone market ensued during the coming years; it's a battle that is still being fought today. But in the summer of 2007, the two companies had come together to launch something brilliant and magical and different out into the world. Something that totally transformed how we find our way—Google Maps for the iPhone.

Chapter 19

GOOGLE'S NEW EYE IN THE SKY

At the start of 2008, Google Maps had been launched in fifty-four countries—and both Google Maps and Earth had tens of millions of monthly users. Daniel now headed a team of business development professionals scouring the globe for mapping data, either to license or buy outright. Larry, Sergey, and Eric made their message to the team clear: Go faster. There was an insatiable appetite among users globally for Google's mapping products.

Soon whole new categories of mapping data began making their way into Google Maps and Earth. Transit data, showing everything from light rail to subway lines to bus stops (often with scheduled departure times), began to appear in Google Maps, thanks to the work of a longtime Keyholer named Jessica Wei. In 2008, a large Google team based in Zurich, Switzerland, was dedicated to building out the transit data feed, often including real-time locations of trains and buses. Google Maps navigation soon began offering walking directions and transit directions, critical differentiators in dense urban cities where getting around is not beholden to car travel.

Daniel's team also began acquiring historical aerial imagery and populating Google Earth with imagery that allowed the user to travel back in time. Brian hired a friend named Reuel Nash from Texas. He showed up in the Austin office one day and went to work on a somewhat-hidden feature of Google Earth that enables the user to look at aerial and satellite photos from the past. Daniel came across historical satellite imagery of the United States from the USSR. The Cold War–era U.S. Keyhole satellites were prohibited by law from pointing their cameras at the United States, but not the Soviet satellites.

Also in 2008, John oversaw the acquisition of a company called Image America, headed by Kevin Reece. Reece's operation had been the provider of the aerial imagery updates for Google Maps and Google Earth during Hurricane Katrina. By 2008, John was so impressed with Reece's business that he convinced Larry, Sergey, and Eric to buy the company. Suddenly Google had its own fleet of aerial imagery airplanes and Learjets. I have never seen Google state the number of planes it owns publicly, though it has been described by the media as the Google Air Force.

At the Googleplex, Reece began working on innovations in the way aerial imagery and 3-D data were captured from this expanding squadron of planes. Together with Luc Vincent and Brian McClendon, he designed a new cluster of cameras to be mounted on Air Google. These cameras swept back and forth as the planes flew, capturing both aerial imagery and 3-D building data. In this manner, the planes could fly lower, for greater image clarity, while still vacuuming up wide swaths of land. Larry was intimately involved in the design of this camera system, including the custom processing chips. The "push broom" camera system is patented by Google, with Reece as the author. It allows Google to quickly capture massive amounts of high-resolution aerial imagery and 3-D cityscapes; the cluster detects every bump and jostle of the plane, accounting for these movements to pinpoint the precise location of every single pixel.

Besides becoming a critical part of the Google mapping data pipeline, Reece's planes have been deployed for numerous emergencies and natural disasters. Similar to Hurricane Katrina and other catastrophic events, these updated photos swiftly make their way into Google Maps and Earth; and they become the base map for the Google People Finder map that has been deployed to aid in reuniting families with their loved ones after natural disasters.

If this wasn't enough, Daniel was negotiating for the imagery from a soon-to-be-launched earth-imaging satellite: GeoEye-1.

The GeoEye-1 was built in Arizona by a company called GeoEye Inc. Its previous Ikonos satellite was providing satellite imagery for both Yahoo! and Microsoft. By 2007, Yahoo! and Microsoft had added aerial and satellite imagery data to their mapping products in an effort to try and keep up with Google Maps and Earth.

Scheduled to launch in the fall of 2008, the GeoEye-1 promised much higher resolution imagery than either the Ikonos satellite or Digital Globe's QuickBird: half-meter resolution imagery compared to one meter meant that each square image tile contained about four times the data. It is interesting to note that both satellites were capable of higher resolution imagery, but the highest resolution images were still reserved for GeoEye's other investors in the launch: namely, the U.S. military.

At half-meter resolution, the GeoEye-1 data was almost

guaranteed to be of excellent quality. And similar to the Quick-Bird satellite, it offered vast amounts of international coverage.

A bidding war between Google and Microsoft transpired to see who wanted the imagery the most. In the end, Google won the multiyear rights to all GeoEye-1 imagery. Because of the size of the deal, Daniel Lederman got scheduled into Larry, Sergey, and Eric's formal deal review meeting. After waiting for three other partnership discussions to wrap up, Daniel presented the GeoEye opportunity to the Google execs, expecting the discussion to be a lengthy one. It took Daniel six minutes to get approval. It was yet another illustration of how Google continued to pour energy and resources into moonshot mapping projects. Eric summed up the sentiment by simply saying, "We can't *not* win this deal." He was forever paranoid of the potential competition from Microsoft.

Now this is something you should know the next time it comes up. When you invest in an earth imaging satellite, it comes with perks.

First, you get to put your company's logo on the satellite, or on the rocket booster, to be precise. Second, you get to watch it launch, and hope that it doesn't blow up on the tarmac (which had happened to Digital Globe).

Vandenberg, the ultra-secret Southern California Air Force base nestled on the Pacific Coast north of Los Angeles, is the launch pad for almost all GPS, military, communication, and surveillance satellites. This makes the scheduling of a rocket launch a difficult task, and short windows of opportunity opened and closed throughout the spring and summer of 2008. Finally, the date of Saturday, September 6, was locked down for the launch of the GeoEye-1 satellite. A select few invitations were extended for the event. Military brass would be on hand, watching with keen interest. GeoEye executives would, of course, be able to attend the launch; and a small contingent from Google was invited to watch the liftoff live. John made the invite list, but had a family commitment and couldn't make it. Daniel was invited and attended with his son Alex.

And then there were Larry and Sergey. GeoEye executives, and even the military leaders, were interested in having their strange bedfellow financial partners join them for the big day. By 2008, Larry and Sergey were full-fledged celebrities in the world of technology, and the Google investment in the satellite was seen as an endorsement of the GeoEye-1 project—a blending of both money *and* credibility.

It was Daniel's job to try to get them there, but Larry and Sergey were notoriously noncommittal. Calls and emails went unanswered. When responses came, they kept changing. One day they were coming. The next they weren't. Then they didn't know. On Friday, September 5, the day before the launch, Daniel answered a call from Larry's admin (Larry and Sergey had finally relented to the idea of an executive admin). "Good news," she said. "Larry and Sergey want to come to the launch at Vandenberg."

"Great," Daniel exclaimed.

"But here's the thing," she continued. "They are wondering if it's okay if they land their jet at Vandenberg. They really don't want to have to land in Burbank and drive."

Daniel was silent for minute. Surely he had misheard what she just said. Of course he knew of Larry and Sergey's three jets that they kept at the NASA Ames airport next door to the Googleplex. The Boeing 767 that the Google founders had bought even caused a boardroom kerfuffle between the two as they argued about whether it should be outfitted with king-size beds or hammocks.

"Wait, what did you say? On Vandenberg Air Force Base?" Daniel repeated.

"Yes. On the Air Force base," she said. "They'd like to be able to fly down on their jet on Saturday morning. Land at the base. Watch the launch. And then just get back on the plane."

It all sounded reasonable, other than the part about landing a private jet on one of the most locked-down epicenters of the military-industrial complex. A base that rerouted Highway 101 so that no vehicles could come within fifteen miles of the launch sites. An active military base shrouded in secrecy on the very day that it was set to launch a $1 billion surveillance satellite.

"Are you fucking kidding me?" Daniel asked, having already spent hours getting the necessary documents and credentials for himself and his son in order to be escorted onto the base in his car.

"No, I am not kidding," she said. "Can you ask? It's the only way they can make it work for their schedule."

And so he did. Daniel called his counterpart at GeoEye.

"Are you fucking kidding me?" the Geo rep responded in disbelief.

Somehow they made it happen.

On the morning of Saturday, September 6, one of Larry and Sergey's jets (they opted for the more practical Gulfstream V) landed on the tarmac of Vandenberg Air Force Base. The Air

Force base's commanding officer, in full uniform, was there to greet them, as were executives from GeoEye. Larry and Sergey emerged from their jet, with Larry's fiancé and Sergey's very pregnant wife in tow. Sergey wore a T-shirt, cargo pants, and bright red Crocs.

They received a behind-the-scenes VIP tour of the operation. The satellite lifted off at 2:18 P.M. By three, Larry and Sergey were airborne, off to their next adventure.

The first image from GeoEye-1 was successfully sent back to Google in October of 2008. It featured the campus of Kutztown University in rural eastern Pennsylvania. GeoEye-1 quickly became a critical source of mapping data for Google, especially internationally. It would be an ongoing and huge competitive advantage for Google and its mapping products.

By the end of 2008, the Google Geo team was almost unrecognizable to me. There were so many new people on the team that I was probably unrecognizable to them, too. Nevertheless, I organized a reunion for the Keyhole team since we had been at Google for four years now.

With a little budget from John, I rented out the NOLA restaurant again, the very room where we had celebrated the acquisition four years earlier. I ordered down vests with both the Google and Keyhole logos for everyone on the team to celebrate that all of our new-employee Google stock options that we received when we started were now "fully vested."

Led by John and Brian, the Geo team now employed over 1,200 employees. There were hundreds of Google Street View cars driving millions of miles of roads across the globe. A fleet of Google airplanes with patented camera clusters captured high-resolution aerial images. Google owned the preferred rights to imagery from two Earth-orbiting satellites. We had added historical imagery. We had added underwater data to map the ocean; now you could dive underwater in Google Earth to explore the terrain of the ocean floor. Transit data flowed in, showing real-time departure and arrival times of trains and buses. We added indoor maps, including going inside businesses with Street View cameras and navigating users through airport terminals and museums. And users were accessing these groundbreaking Google mapping products multiple times per day on new mobile devices to find their way in the world.

The pace of the Geo team made my head spin. Every time I caught up with John, he appeared to be both exhausted and exhilarated. By the end of 2008, Google Earth had been downloaded over five hundred million times.

Chapter 20

PROJECT GROUND TRUTH

As you can imagine, from 2005 to 2008, data providers were more than pleased with the unprecedented popularity of Google Maps and Google Earth. It meant that contract renewals often started by adding a zero or two to the prior contract's price point. By 2008, John had a "success failure" problem that threatened Google Maps's very existence. The economic viability of relying on other companies' mapping data to create Google Maps and Google Earth was questionable.

While usage of Google Maps and Earth on desktop computers grew through 2007, the *real* problem child starting in 2008 was the widespread usage of Google Maps on mobile devices—both on iOS and Android.

It had taken Google two years to ship its first Android: The HTC Dream debuted in November of 2008. It was a decent device, especially if you didn't compare it to an iPhone. But the Android operating system's true allure was the fact that it was free to the wireless carriers and smartphone device manufacturers, *and* it allowed them to create their own customized versions. With this open and customizable approach, Android took off and soon outpaced other mobile operating systems. Today, Android and iPhone make up 99 percent of all smartphones sold.

As Google's mapping products exploded in popularity, the costs of licensing data also escalated dramatically. The most egregious price gougers were the road-network data providers. This industry had consolidated to two major players: Tele Atlas out of the Netherlands and Navteq based in the United States. The industry had once been comprised of dozens of regional players, each creating databases of the roads in a geographic region. Their process included hiring hundreds of mapmakers to look at aerial imagery and enter data about each street, including attributes, such as speed limit, one-way or two-way, number of fire hydrants, and other details. All this road-network data worldwide had been snapped up by two huge and suddenly powerful companies.

By 2008, these two companies wielded immense power over John and all of Google's mapping innovative ideas. They dictated not only how much we were to pay each year but also how we could and couldn't use their data. Navteq and Tele Atlas didn't want to cannibalize in any way their bread-and-butter navigation device market. They were beholden to the Garmins, TomToms, and Magellans of the world, whose sales of personal navigation devices numbered in the billions. If Google Maps were allowed to have turn-by-turn directions with voice-over, they feared (with good reason) that it might destroy that market.

The pricing structure of Navteq and Tele Atlas, which carried over from early web mapping contracts, became economically untenable in a world with Google Maps everywhere. These contracts charged Google based on something called mapviews, which were defined as the number of times that visitors viewed a map screen that used that company's data. For Google, with its fast and fluid maps, users were viewing exponentially more mapviews in Google Maps and Google Earth than they were on slower-to-load sites like MapQuest. It was like the difference between viewing a single index card versus thumbing through a flip-book of index cards: On Google Earth, you might flip through a hundred cards in ten seconds, and Google was financially on the hook for every card viewed.

In the face of rapidly ballooning costs in 2008, Daniel and John went back to Navteq and asked instead for a flat-fee deal. Within forty-eight hours, Navteq came up with their exorbitant price—*without* spoken turn-by-turn directions.

John and Daniel toyed with the idea of buying Tele Atlas. They met with the company's investment bankers, but the price tag soared into the billions, and the data itself wasn't "clean," meaning it had been assembled from a geographically dispersed set of acquisitions with a variety of usage restrictions.

Against this unsustainable economic backdrop, John, Brian, Daniel, and the team began to consider an outrageous alternative.

If you recall, the Google Street View project began as an experiment in 2002, with Larry, Marissa, and Sergey out for a Saturday drive. In 2008, Luc Vincent's team went off-road with their technology, taking Street View cameras into new, unexpected places. First, a tricycle was equipped with a Street View camera, and park paths and trails were mapped. Then a cart was mounted with a Street View camera, and a variety of museums and other indoor buildings were mapped. Then a snowmobile was equipped with a Street View camera, and ski slopes were mapped. For extreme downhill ski runs, Vincent strapped a camera to his back and attempted a not-so-unsuccessful run down an advanced slope at Squaw Valley. As Vincent struggled in the thick powder to keep the robotic 80 MB spheroid of five synchronized cameras upright and not sink into the snow under its weight, a teenager skied up and asked, "Hey, dude, why don't you just get a GoPro?"

By 2009, Google's fleet of Street View cars had traveled over thirteen million miles in twenty-two different countries. It was a stunning feat of technical ingenuity and created delight for hundreds of millions of Google Maps users worldwide. In my opinion, however, there was no chance that it would ever make any kind of economic sense for Google.

I was dead wrong.

That's because in early 2008, Sebastian Thrun had a new idea for the use of Google Street View data: one that might fundamentally shift the economic foundation on which Google's mapping services were built. In 2008, Thrun walked into Brian McClendon's office and said, "I think we might be able to make our own road-network maps out of the Street View data."

His idea, he explained, would require multiple thousands of people. But it was an idea that also could flip the impact of the Google Street View project. Thrun's idea, if it worked, would transform Street View from a team that cost Google tens of millions of dollars into one that might save the company hundreds of millions, if not billions, of dollars.

Replacing the map data being licensed, at ever-increasing prices, from Navteq and Tele Atlas, would save Google billions. And price wasn't the only reason that Navteq and Tele Atlas maps weren't going to work for Thrun's group. There was also the matter of the quality of the road-network data and the speed at which data providers would update.

I'm sure you've experienced this yourself. I know I have. Some new road opens up or some street closes, and your navigation devices aren't current with the latest and greatest ground-truth information. In a world based on maps drawn using data licensed from third parties, if a user reported a problem with the navigation, saying that something had changed, it kicked off a long sequence of events for that problem to get actually corrected. First, that report would go back to the data provider. That data provider would attempt to verify the change and then would decide whether or not to update their road network database. Assuming the change was made, it would be sent out in the next database update delivered to Google; by the way, this update would also be shared with all of this provider's other subscribers. In the best-case scenario, the reported problem might make its way to an updated Google Maps or Earth client within six months, though it was more common for that process to take over a year.

This kind of workflow was not viable for Thrun and Levandowski's self-driving cars, which needed incredibly accurate and current maps. (They needed maps that were updated daily, not annually.) By this time, Thrun and Levandowski's self-driving car group at Google employed about ninety people, and they were already testing prototypes of the cars in the vast parking lot of the Shoreline Amphitheatre. I had seen the cars in action on trips back to Mountain View, zipping around orange pylons through the dusty unpaved parking lots between the Googleplex and the Google offices on Crittenden Avenue. It was a top secret project; I mistakenly thought they were the next version of Google's Street View cars (their performance and safety were still being monitored by a Googler in the driver's seat), when in fact they were autonomously driving over one hundred thousand miles on California roads.

The self-driving car group was completely separate from the Google Street View team, operating independent from the rest of the Geo team. But the group did use some of the Street View computer vision technology to recognize street signs, speed limits, and addresses as its cars began to cruise the streets around Mountain View.

The idea of using Street View imagery to extract data wasn't without precedent. In late 2007, Google began using Street View imagery to verify needed updates to business listing information. If a business owner reported that his or her business was in the wrong location, the Google Street View imagery would be used to verify the reported issue, and then the location of the business was corrected.

Thrun proposed to take this idea much further. He lobbied Brian and John to consider founding a new project: to use Google Street View imagery and computer vision technology to annotate the entire planet, to extract road-network data from the images captured. It would be the start of Google's most ambitious mapping project ever, dwarfing even Google Street View—and the whole Google Geo team operation, for that matter—in its scope, budget, head count, and complexity. It's a project that has gotten very little attention outside the walls of the Googleplex. Even today, internally at Google, it is known as one of the most outrageous, moonshot projects ever undertaken by the company.

It is simply called Ground Truth.

The Ground Truth project was the fruit of Larry Page's original vision on his Saturday drive back in 2002—using street-level images to index the *physical* world.

John went to Larry with the idea to fully fund Ground Truth. It was a massive commitment, not one to be taken lightly. It was going to require hundreds of software engineers and product managers, and thousand and thousands of map drawers—or Ground Truth operators, as they would come to be known—who would redraw the entire planet on a blank sheet of paper. Once Google took this path, it couldn't turn back: There was no returning to the data providers once the decision had been made.

For these reasons, John asked Larry for a five-year commitment to the project. There was no way that the project should be started, he argued, if they had to revisit the initiative's budget every year. In the summer of 2008, Larry gave John and Brian the go-ahead, and Ground Truth was founded.

The first step for Ground Truth was to develop a brand-new map-drawing software tool called Atlas, which was a sort of advanced cousin of Esri, customized for the Google Geo team's unique needs and assets. Atlas is a remarkably sophisticated, complex, and smart map-drawing tool. When I first saw a demo, it struck me as a kind of mash-up of Google Earth, Street View, and Adobe Illustrator; it allowed the user to draw lines and add annotations.

Atlas imported all Google's mapping data for a location into a single view. The aerial and satellite imagery processed by Wayne Thai's team always provided the foundational layer. In other words, it was the sheet of paper on which the roads were traced. On top of that imported imagery, Atlas arranged thousands of dots, each one representing Vincent's team's Street View snapshots. For the next layer, Atlas accessed any freely available road-network database from a government source. In the United States, this data set was called TIGER (Topologically Integrated Geographic Encoding and Referencing), and was created by the U.S. Census Bureau and therefore a part of the public domain (read: free).

That said, TIGER's data set was of an inferior quality, geographically speaking. Roads were notorious for being poorly aligned to their actual locations, since the Census Bureau uses the data only to count households. As imperfectly positioned as these lines were, Atlas still started with them so that the Ground Truth operators would not have to draw from scratch. This made it easy to simply drag a road over to its correct location, using the aerial image below as the Ground Truth location of each road segment.

Atlas also featured a fish-eye view for each road segment. When you rolled a cursor over the thousands of dots on any road, the most recent Google Street View image panorama would immediately pop up, taking the Ground Truth operator into the immersive and most current view of the street.

All imagery was automatically processed through Google's computer vision algorithms, meaning that dozens of pieces of data about each road segment were automatically extracted from street signs and address markers. Everything from speed, school zones, number of lanes, left- or right-turn restrictions, and a dozen other metadata features were magically pinned to every individual road segment. For example, Atlas would look at an aerial image that showed all the cars on both sides of the street pointing in the same direction and automatically deduce that it was a one-way street. It was as if the software had its own mapmaking brain.

As good as Atlas was, however, it was only queuing up the data to be analyzed by a human: A Ground Truth operator was still needed to review, edit, and verify what Atlas purported to see. For example, Atlas would not automatically assign a street to be, say, one-way; instead, it highlighted the street and suggested that it was one-way. A Ground Truth operator would go in and open the Street View imagery of that road before confirming that, yes, in fact, Atlas's computer vision was correct. The street was one-way. Only after it was reviewed by a human would the data go live into Google Maps and Earth. At the end of the day, this meant that a massive amount of human elbow grease was required to create a complete, accurate road network database of the entire planet.

While Atlas did give operators the most state-of-the-art, computer-vision-enhanced, map-drawing tool, the process was similar to the workflow of other road-network data creators. I happened to observe such an operation firsthand back in 2002. John dispatched me from Boston to go visit a company called Geographic Data Technology (GDT), which was located on the outskirts of Lebanon, New Hampshire. It didn't look like the most stimulating work. About fifty people stared at computer monitors all day, examining aerial images and using printed maps to draw digital road segments in the right location. As part of the industry's consolidation, GDT and all their data were acquired by Tele Atlas.

Dozens of these types of map-drawing companies had existed for years, with teams of people drawing millions of miles of roads across the planet. Google and the Ground Truth team proposed to do it all from square one—in two years. The clock was ticking on the next round of contract negotiations with Tele Atlas and Navteq.

With a multiyear financial commitment from Larry and a beta version of the Atlas tool ready for use, the first Ground Truth operators secretly set up shop in Building 45. The facility was unlike any other at Google: Rows and rows and rows of computers were arranged right next to one another on simple white tables with picnic benches (instead of chairs). Forget not getting an office. These guys weren't even given a chair.

The spare surroundings were for legal reasons: Google wanted to be sure that the data generated by Ground Truth operators was completely clean and uncorrupted from any other source of external mapping data. All the computers needed to be completely visible to the Ground Truth operator supervisors, and all the computers running the Atlas tool could run nothing but Atlas. All other websites were blocked. The Ground Truth data needed to be clean and wholly created by Google. (These employees also were prohibited from bringing in their cell phones in order to maintain the data purity.)

The project in Building 45 quickly ramped up to two hundred Ground Truth operators, then five hundred, with more computers and more benches crammed into the hallways. Multiple eight-hour shifts occurred throughout the day. Then the enterprise expanded overseas to two thousand Ground Truth operators, then to five thousand.

In the summer of 2009, despite the tremendous progress of the Ground Truth team, John and Daniel knew that Google wasn't going to be ready to switch away from the existing mapping providers anytime soon. Even as countries became available, we would need deals in place to cover Google where our Ground Truth mapping data was not yet complete. This ambitious project needed more runway; it was up to Daniel and John to negotiate more time. As a result, John and Daniel began the arduous process of negotiating renewals with Navteq and Tele Atlas for what they both hoped would be the very last time.

In 2009, Daniel, John, and Vikram Grover (from Daniel's team) flew to London to meet with Tele Atlas in the Google office. There they met with Tele Atlas CEO Alain De Taeye, account director John Sheridan, and the executive director of its Location Based Services, David Nevin.

They started the meeting by listing out the rapid growth of Google Maps on mobile devices. Daniel later told me that their message was essentially this: "If Google thought for a minute that the price was going to be anywhere near what was paid last year, then we had another thing coming." In fact, the meeting almost ended there, with John threatening to get back on a plane bound for San Francisco.

Over the next two months, Google and Tele Atlas met dozens of times and exchanged over ten contracts in order to hammer out a two-year deal. What Tele Atlas couldn't have known was that their pricing scheme had essentially seeded Google's efforts to map the world on our own. Ground Truth was in full production by then. In retrospect, Daniel thinks that Tele Atlas might have suspected something was in the works. In several iterations of the contract, Tele Atlas tried to insert provisions precluding Google from building its own mapping data sets. But Daniel and

John could not—and would not—agree to these last-minute provisions. They argued that Google needed the ability to create its own data sets because there were places that Tele Atlas might not cover, such as underdeveloped and remote locations (which was at least part of the truth).

By the fall of 2009, after months of negotiations, the deal was almost ready to be signed. John, Daniel, and Vikram flew to Zurich, Switzerland, to finalize the negotiations and attend a celebratory dinner. The Zurich Google office, like all Google offices, had its own architecturally whimsical features, including a fireman's pole that allowed Googlers and visitors to slide down from the fifth floor to the fourth floor. John, Vikram, and Daniel frequently used the pole on their way down to lunch; even the Tele Atlas CEO would take the pole. But the Tele Atlas head of marketing, who was a frequent participant in the negotiations, refused to take the pole. He was a tall and stocky guy, and was reluctant to try it.

But as the negotiations dragged on, one day after a fair bit of razzing from the others, their head of marketing said, "I'll take the fireman's pole down when we get the deal signed." It was a running joke between the teams and served as a good-natured incentive to get the deal completed.

And now it was done. On their way out to the celebratory deal-signing dinner, Daniel, John, Vikram, and Alain all took the pole down and assembled at the bottom waiting for the head of marketing. Wearing a heavy backpack, he stepped up to the pole, grabbed it with two hands, and jumped on—and then dropped like a stone. Forgetting to wrap his legs around it, he plummeted to the fourth floor, crashing into a heap at his boss's feet, breaking his leg in the process.

Six weeks later, the cast had just been removed when the head of marketing and De Taeye visited Daniel and John in Mountain View for a follow-up meeting, which likely became the worst meeting of their careers. Though Google negotiated the rights to use the Tele Atlas data, we were not obligated to use it (even though Google would still pay for the use of that data for the remainder of the contract). In fact, thanks to the Ground Truth project, Google was ready to swap out Tele Atlas data for its own. Year after year, Tele Atlas had raised the stakes—and the pot had grown to be worth billions of dollars, putting into question the economics of Google mapping products and limiting the company's ability (for example, no voice turn-by-turn directions while using their data). John described the meeting "like playing poker and knowing you have a royal flush." Tele Atlas created the economic incentive for Google to pursue a radical alternative: to start an outrageously ambitious technical and logistical moonshot effort to redraw the entire planet's road network on their own. And the first data set from Project Ground Truth—a data set of all roads in the United States and Mexico—was ready. John informed De Taeye that Google would be turning Project Ground Truth data on for all Google Maps and Google Earth users in three days. Google Maps and Earth would no longer use Tele Atlas road-network data in the United States and Mexico, and by the end of the year, Google would no longer need their data for anywhere else on the planet.

All the mapping had been created by Google—free of all royalties and all restrictions. Google would be introducing voice-over, turn-by-turn directions free on Google Maps for mobile products. First on Android, and then on iPhone in the near future.

Tele Atlas was out. Ground Truth was in.

Today, Ground Truth is the foundational data layer for all of Google's mapping efforts and self-driving car initiative. If you have ever used Google Maps and had it speak out directions, you have Project Ground Truth to thank.

Since 2009, a web-based service that emulates the functionality of the Atlas tool has been published by Google; it's called Map Maker, allowing users in countries that have notoriously poor road network data to self-map their area. (Map Maker is now a feature of Google Maps.) That data now also feeds into the Ground Truth data set just like the data created by the Ground Truth Operators using Atlas. Many countries have self-mapped themselves using Map Maker. For example, India, with a population of 1.2 billion, has been completely mapped through Map Maker.

The concept of relying on users with a set of tools and processes to draw the road networks of the world took off as an independent open-source initiative in the late 2000s. Inspired by Wikipedia, OpenStreetMap (or OSM) is a collaborative effort started in the UK to create free editable maps of the entire world. It has over one million local contributors and its coverage today compares favorably to proprietary providers of mapping data, including Google.

Ground Truth processes and data are now the underpinnings of Google's self-driving car initiatives, too. For the self-driving car project, now a spin-off company called Waymo, having highly accurate, highly current data is critical. If a new road has opened or a road has been closed, Waymo needs to know as soon as possible or the whole self-driving car initiative could wind up being a dead end.

Google wouldn't have to wait for a data provider to edit and update its database six months later. Today, thousands of updates per day flow into the Ground Truth team via the "Report a Problem" link in all Google mapping products. Assuming that the same problem is being reported by multiple users, that issue gets queued up for a Ground Truth operator to take the case and review it. With the help of thousands of Ground Truth operators, that list of issues is maintained with zero backlog: Reported problems can be taken care of within minutes and maps can be updated across all Google mapping products instantaneously. Not only that, if there are several reports of a missing address, Google takes that as a hint of a new subdivision, triggering Kevin Reece to redirect one of Google's aerial image-capturing Cessnas to fly over the area and update it. A Google Street View car is similarly dispatched to map the new roads.

Today, Google's mapping efforts are moving toward creating a dynamic real-time monitoring of the planet. The acquisition of Waze, for $966 million, allowed Google to pull real-time traffic incidents and data into all its mapping products. Now the directions you receive on Google Maps are dynamic and include traffic accidents and backups. The acquisition of Skybox Imaging, for $500 million, allowed Google to launch its own Earth-monitoring satellites, though Google has never reported on the number of satellites it has launched.

When I joked with the real estate guys at the trade show to go outside and wave hi, it was funny because the idea was preposterous. Now I'm not so sure. Before it was bought, Skybox offered retailers and investors the promise of predicting store sales of competitors by monitoring the number of cars in the parking lots. From space. For example, Home Depot could use Skybox satellite imagery to predict the sales trends of any given Lowe's hardware store and plan to locate a new Home Depot near a Lowe's store that appeared to be trending upward.

Our planet is a dynamic place. And today, Google Maps and Earth are trying to keep pace.

Chapter 21

MOONSHOT COMPLETE. NOW ON TO MARS.

Those days, in 2010 and 2011, I traveled from Austin to the Googleplex every two months or so, and always made a point to stop by and see my Keyhole and Google Geo friends for lunch or happy hour. (By this time, the Geo team, numbering about seven thousand people, including the Ground Truth operators, had its own building on the campus.) Sometimes John and I went for a run along the familiar trails of the nearby Shoreline landfill, took in an Oakland A's game, or met up for a beer at the Sports Page.

On a beautiful day in September of 2010, we met at NOLA in downtown Palo Alto. It had been several months since we had last seen each other, and when John walked in, he looked beat down as he slipped into a chair at the bar. Despite all the successes, awards, and promotions, he looked spent. The turf battles between Brian and him over various product issues had been ongoing, and he looked like a boxer who had endured nine rounds of body blows. For example, Germany had recently fined the Google Maps team for accessing prohibited Wi-Fi networks while roaming German streets. He admitted that he felt like he needed a change of scenery.

"You aren't thinking about leaving the company?" I asked.

"We'll see," he said. I thought that was the end of the conversation. John shifted in his chair and looked away. I read his body language to mean that he wanted to move on to another subject, but this time I was wrong. "I don't want to be just a one-hit wonder," John continued. "I don't want to be the guy who did Google Maps and Google Earth."

I took a swig of my beer, thinking about the hundreds of millions of users of Google Maps and Google Earth, and reminded him, "Well, John, it sure was one *hell* of a hit."

John looked back at me from the Oakland A's game on the television mounted above the bar. "I feel like I want to swing the bat again. But the start-up thing is hard. I mean, think about Keyhole and doing that all over again. If I'm going to do it, I've got to do it soon. It's a young man's game, and if I'm going to get back up at the plate, I want to do it while I've still got some pop in my bat. You know what I mean?"

For all of John's successes and all of the accolades for Google Maps and Google Earth, I witnessed only fleeting moments of John's being satisfied. Someone once described him to me this way: "It's like we have been on this incredible journey to the moon, and after all the struggles, scratching and crawling to find our way, we have finally made it to the moon. And as soon as we arrived to the moon, we have barely caught our breath, and our leader has said, 'Okay, ready? Now we must go to Mars.'"

"Would you consider doing a start-up again?" John asked me pointedly. He knew that my job—though strategic for Google—wasn't fully tapping my creative marketing side.

I could tell he was anxious. Was he asking me if I was interested in a job? *A job outside of Google?* Suddenly I got the sense that something was up. Something imminent. Was he actually probing to see if I'd consider leaving Google? Leave the cozy confines of the Googleplex, with its perks, food, cachet, paychecks that didn't bounce, and copiers that always had paper in them? Criminy, at Google there was even an employee who roamed the offices ensuring that the colorful exercise balls strewn around campus were properly inflated. Would any new start-up have an exercise ball inflation coordinator? I still said a little novena to St. AdWords every time I put my badge up to a Google office badge reader, praying the company would let me work there just one more day.

"Wait, are you asking in the abstract here, Hanke?" I asked. "Because, man, to be totally honest, for me personally, it would be hard. There's the Google stock options. And, I mean, I'm not like you, John. I'm perfectly happy being a one-hit wonder." And then I took a sip of my beer and added for emphasis, "It's a lot better than being a *no-hit* wonder."

"Yeah, I guess it is," John said with a weak chuckle. But I could tell that he was disappointed in my answer. And I got the sense that he did in fact have a card up his sleeve. My tepid reaction to the idea of leaving Google to create another start-up wasn't what he was hoping to hear. An uncomfortable silence settled in the air.

In October of 2010, Larry Page sent out an email to all@google.com announcing a reorganization and shuffling of executive responsibilities. The lead engineer of Google search, Udi Manber, was being promoted to be the head of product management and engineering. This meant he was inheriting Marissa's search team responsibilities. Marissa, Page announced, was going to be returning to one of her early passions at Google: maps and location services, otherwise known as Google Geo.

Uh-oh, I thought to myself as I read the email in the Google Austin office. *This is not good for John.* I waited a few days before calling him.

"So, Marissa? What the hell?" I asked. The move had caught John by surprise, too. He told me how it had all gone down.

John was at an off-site in New York City with Elizabeth Harmon, Dan Egnor, and the rest of the Geo leadership team in Google's sprawling offices in Chelsea. As a team-building exercise, they had played an urban treasure hunt called the GO Game, and afterward the group went and saw *The Social Network*. After the movie, John checked his phone and noticed that he had received a call from Marissa Mayer. By this point, six years into Google, John had come to respect and like Marissa. John and Holly had even been invited to numerous social engagements, including Marissa's recent wedding. They had become friends.

But the voice mail was odd. Marissa said she was *in New York City* and wanted to meet the next morning at the Google NYC offices. The following morning Marissa told John what was going to happen: There was a major shake-up at the top of Google's search org; Udi Manber was consolidating power and she was getting shifted around. While Geo had always been an interest, John and Brian's world had occupied only about 5 percent of her focus. Now, she explained, it would occupy 100 percent of her attention. Marissa was taking over Google Geo. She explained that John and Brian would be reporting to her. "I want you to stay," she said. "I'm hoping we can all make it work."

As they spoke that morning, the wheels for John's next game plan were already turning. "That sounds great," he said to Marissa, but in his mind, he thought, *This is my time to move in a new direction.*

Two weeks later, Marissa gathered the entire Geo team for the first time in Mountain View and officially introduced herself to her new team. During this meeting, she also announced that John was leaving Geo. He planned to stay at Google to work on an as-yet-unannounced project. It was a project he was excited about and had been thinking about for a while.

For about six weeks, John worked out of the Google office in San Francisco. Commuting from Oakland on the San Francisco Bay Ferry, he thought long and hard about his next swing of the bat. Eventually he decided to leave Google altogether, with the idea of creating a new start-up based on Google's mapping technologies. When he told Larry that he was leaving and was going to establish a new start-up, Larry asked him what the idea was.

When John told him the concept, Larry said, "Well, why don't you just stay at Google and create the start-up inside Google?" More than anyone, Larry had been the driving force behind Google's bold mapping moonshot projects. He continued to invest in efforts to map and index the entire physical world, and he hated to see John leave. He knew well of John's vision, determination, and drive. If John wanted to create something new, whatever it was, Larry supported him, just as he had at every turn of the Google Maps and Google Earth journey.

Together they worked out a deal on a single sheet of paper.

It was a novel arrangement. Employees in a new company, called Niantic Labs, could forgo their Google stock in exchange for Niantic equity. There was a three-year runway, at the end of which the new company's equity would be paid out based on the valuation of this new company. Technically all employees were still Google employees, but John would be free to run the group as if it were its own separate entity. It was the best of both worlds: the freedom and upside equity of a start-up; the work environment and perks of Google.

John's concept for the new company was born from an ongoing battle with his son: the all-too-common fight that parents have with teenagers sucked into the allure of video games, even on beautiful California weekends. "Go outside and play!" John would say often to Evan, cutting him off from Minecraft or some other video game that kept him stuck at home staring at a screen. Eventually father and son agreed on a deal. Evan would be allowed an hour of gaming for every hour he spent outside. It would become a kind of mission statement for Niantic, marrying John's knowledge of maps with his passion for games, creating apps that once again got people—young and old—to go outside and play.

The idea was to use mobile phones to turn maps into games. To create mobile app games that, in order for you to progress in the game, forced you up off the couch and into the real world, and to augment that reality with game pieces mapped to the physical world. "Real-world games," he would call them. The goal is to get people moving out from behind their screens and appreciating the world around them in new ways.

John named his new company Niantic, which had been the name of a merchant marine vessel that landed in San Francisco in 1849, carrying 246 fortune-seeking men who immediately rushed from the ship, leaving the *Niantic* behind with dozens of other ships. The *Niantic* is buried underneath modern-day San Francisco, directly under the Transamerica Pyramid, though few passersby bother to notice or care. John hoped these new GPS-based (or location-based) services and games could help highlight the hidden history all around us, thus enabling us to appreciate locations in new ways. If you play a game from Niantic, you end up learning a bit about the world around you, whether you meant to or not.

I was not alone in following John to Niantic. All told, eight former Keyholers joined John's new start-up. One day Lenette emailed me: "Here we go again!" As has often been the case throughout my professional career, I can't say that I fully understood the concept that John wanted to create. I was happy that John asked me, of all the marketing people he has worked with, to be the lead marketing guy for Niantic. And when someone questioned me about leaving Google proper to join John's upstart team, I told her, "Look, if Michael Jordan asks you to be on his team, you just do it, even if you don't know what sport is being played."

It is with a supreme sense of wonderment that—as I write this thirty-two years after meeting him—I am still working for John Vincent Hanke: that same small-town West Texas kid with whom I happened to be randomly assigned to live on the same hall in the largest dorm in the world. My mother has been known to tell it this way, often in the presence of friends and family. "Well, Bill Kilday, your whole career is basically based on meeting one guy."

She's right. On the western frontier, the old motto of the Texas

Rangers was "One Riot, One Ranger." Mine might be "One Guy, One Career." I can admit it. I'm sure that Steve Jobs had his cadre of marketing people behind him too, following him around from project to project like loyal foot soldiers. Ask them if they care about attaching their career to one person: You can probably find them sitting on a beach with a mai tai.

In October of 2014, I organized another Keyhole reunion, this time to celebrate the *ten-year* anniversary of the acquisition by Google. The team had largely scattered. Sadly, Andria Ruben passed away in 2010; Daniel Lederman and David Kornmann lived overseas; and several more were out of state. But of the twenty-eight living members of the Keyhole team, twenty-four showed up for the reunion.

John was the last to commit. "You know I'm not one to live in the past," he told me when I finally cornered him. "John, you are coming. Not coming is *not* an option. You have to be there," I demanded. I think the fact that Dede Kettman was driving with her husband in an RV from Arizona finally made him commit.

Brian, an avid photographer, provided both the slide show and the budget for the event. After John left the crowded Google Geo team in mid-2011, Marissa left soon thereafter, in 2012, to become CEO of Yahoo! This meant that from 2012 to 2014, Brian ran the seven-thousand-person Google Geo operation.

That period had included presiding over Google Maps during the time that Apple dumped Google Maps in favor of its own maps. Apple Maps had been a disastrous launch for Apple, resulting in the loss of $30 billion of market capitalization and a public apology from CEO Tim Cook, and it cost Scott Forstall, head of Apple's iOS, his job.

The day of the reunion, October 16, 2014, ten years to the

day from the acquisition, Brian resigned from his position and passed the baton of Google Geo to longtime Googler Jen Fitzpatrick. Within months, Brian left Google to run the self-driving car initiative at Uber (that's another story).

When I asked Brian why he decided to leave Google, he said, "It was ten magical years, and it was just time." Google had naturally evolved to be a more mature company, with more conventional business practices. There was a new CFO in town, and Ruth Porat was applying traditional business criteria to projects. Basic questions were being asked about the long-term economic viability of all sorts of moonshots Larry Page had started. Google Maps and Google Earth had been like an unsupervised ten-year pillow fight with a teenage babysitter, and now adult supervision had just walked in the door.

Ten years earlier, on our very first day at Google in 2004, Michael Jones suggested to Sergey that one day the Google Earth team might need as much as one petabyte of mapping data. It was the first time that I had heard that number. One million gigabytes. *One petabyte.*

By 2014 Google's mapping product databases represented *twenty-five* petabytes. And they were still growing. The Geo team Brian and John had left behind was an efficient map-publishing machine: Every two weeks it publishes an amount of data greater than the total amount of mapping data the company had at the launch of Google Earth (much of it by Keyholer Wayne Thai and his team).

In that same meeting ten years earlier, I had asked Larry to make a choice between making $10 million or having 10 million users. I was off by a factor of one hundred.

All along the way, Larry and Sergey were consistent in their inconsistency with typical business principles. Given the choice between user delight and money, they went with user delight 100 percent of the time. It's the reason we didn't wind up at my level of 10 million users. Instead, Google Maps and Google Earth wound up at the level of 1 billion users. Monthly.

For me, here's the most amazing thing about the whole ride. The thing that you might find impossible to believe. No one interviewed for this book could tell me the answer to a simple question: At the end of the day, after all the energy and all the data and all the money and all the innovation, have Google Maps and Google Earth actually made money for Google? Have Google Maps and Google Earth turned a profit?

Of course, there is the perspective that Google Maps and Earth lifted the Google brand, provided strategic leverage with companies like Apple, imparted Android devices with superior navigation features, and geographically optimized search results and ads. There's little doubt that it has earned billions and billions of dollars across all Google properties.

But the whole making-money thing. *It's not the primary driver of why they did it.*

Yes, at the time of this writing, a single Google share is at a split adjusted price of $2,000 and Larry and Sergey are ranked, respectively, as the eighth- and ninth-wealthiest people on the planet. But making money is not why they did it. It is not why in 2002 they drove down Highway 101, pointing a camcorder out the window of Larry's car. It is not why they bought Keyhole. Or Where2Tech. Or SketchUp. Or Waze. Or Skybox Imaging. Or Kevin Reece's fleet of planes. Or started Street View. Or started Project Ground Truth.

I'm not saying that it wasn't their top priority; I'm saying it wasn't in their top *ten* priorities. I can promise you, in the meetings I was in, they hardly ever asked about the money. They didn't ask about return on investment. They didn't ask about the payback period of an investment. A conventional company would have been asking all these questions (and more), but as Larry stated in his letter to shareholders, Google was far from a conventional company—and didn't intend to become one. They were interested in one thing: making bold bets on extraordinary products that organized the world's geographic information, and then giving it all away via remarkable products like Google Maps and Google Earth.

I'll give you an example: After the acquisition, Keyhole sales reps Jeff Kanai and Greg Lloyd worked hard to turn the old Keyhole professional license sales (the same business we started by selling individual licenses with an old-school credit card slider at ICSC and other trade shows) into an $80-million-per-year business at Google. And then one day Larry decided to give the Google Earth Pro version away for free, too.

Google Maps and Google Earth have been, as crazy as it sounds, a gift to the world. And for that I say, "Thanks."

Many stories were shared that night of the ten-year Keyhole acquisition reunion. Many toasts were given—from Michael Jones, Brian McClendon, John Hanke, Phil Keslin, Chikai Ohazama, Mark Aubin, and Lenette Posada Howard.

Surprisingly, John was the absolute last person to leave the party. I remember the manager of the restaurant locking the door behind him as we walked out together. John enjoyed himself more than anyone, reconnecting with the whole team, especially seeing Dede, for whom he has such obvious fondness. I don't think he wanted the party and the memories to end. For a single night, John was, I think, allowing himself to live in the past and savor the successes. For one night.

As we stepped out into the crisp fall California night, he patted me on the back. "Hey, man, thanks for putting this together, and thanks for making me come. This was really fun. There were so many great stories I've started to forget about," John said. Unknowingly, he had just given me the nudge I needed on a project I had been considering.

A few months later, John was in Austin for SXSW, the international interactive film, and music festival started thirty years earlier, the very same year John and I arrived at UT. In 1985, several bar owners on Sixth Street and the editor of *The Austin Chronicle* had banded together to give students like John and me a reason to stick around Austin during spring break. Now it is an international megafest, attracting hundreds of thousands of visitors to Austin over nine days every spring. It is an annual pilgrimage for John—for the conference and for his fix of barbecue, a swim at Barton Springs, and live music, and to visit his mom in Cross Plains.

Before he headed out of town, we went for a run around Lady Bird Lake. It was Friday morning, and the trail was crowded. We had to watch carefully to avoid running into a pedestrian with his head buried in his phone, staring at his blue dot representing his location. Runners tracked their routes using MapMyRun, trying to beat their personal bests. Cyclists Strava-ed, trying to beat Lance Armstrong's time on this loop. Commuters were Ubering. Tourists were Yelp-ing and Hotel Tonight–ing. Home buyers were Zillow-ing. Singles were Tinder-ing. Taxi drivers were Waze-ing. Dog owners were Whistle-ing to track their dogs. A UPS truck drove by with packages being tracked and mapped by the recipients. The plane flying overhead was being tracked and mapped by waiting relatives. An unsuspecting Austin High teen's blue dot was being tracked by his mother.

It was a zombie apocalypse of blue dots that John and I had inadvertently helped to create. I took solace in the fact that at least the zombies seemed to know where they were going.

As we crossed over the lake and made the turn onto the four-mile loop, I finally got up the gumption to tell John about a project I had been thinking about since the reunion. A book. This book. I was nervous how he would react, as he was loath to talk about himself and the past. But he was surprisingly positive.

"It's a story that should be told, and you might just be the perfect person to tell it. You were there to witness it all," John said.

When I told John that I planned to end the book in 2006, when I left the Google Geo team, he said, "No, you should write all of it. The whole story. Someone needs to."

Immediately I began peppering him with questions about Street View and Ground Truth. Within half a mile, John had picked up his pace. It was clear to me that when he said, "You should write it," he might as well have said, "You—*not me*—should write it." In case I didn't get the hint, John held up his hand and said flatly, "I don't want to talk about this anymore." We finished the run in silence.

Standing next to his rental car, John threw on a dry T-shirt for his long drive to Cross Plains. "You have to understand," he said, "a lot of what we went through was not all that pleasant for me. The delayed payrolls, the lawsuit, the late vendor payments at Keyhole. And then the turf wars with Bret and Marissa, and product battles internally at Google. The long hours. It was hard on me. It was hard on *Holly*. It's hard for me to put myself back there. Plus, I'd rather think about the future, and what's next. What's coming. Not the past."

It was clear then that if I wrote this book, it would largely be without John's help.

In the spring of 2015, Austin was a big city, with real traffic, major construction projects, and road closures for SXSW. John settled into his rental car. Forgetting who I was talking to for a minute, I asked, "Okay, do you know how to navigate back out to Mopac and onto 183?" Like everyone else, John had focused his attention on his phone.

He waved me off with a smile. "No thanks. I think I got it."

He finished typing *Cross Plains* into Google Maps—and hit *GO*.

EPILOGUE

Did You Get One?

It is Sunday, July 17, 2016. I am in Tokyo, Japan, downstairs from the street level in a tiny, darkened Japanese steakhouse, sitting in a draped-off room just barely big enough for me, John, and his son, Evan. Evan is headed off to New York University in a month, and John is taking him on a one-week father-son trip across Japan.

The restaurant's owner is explaining how to cook our meats and vegetables over the ishiyaki grilling stones in the middle of the table. The owner is here because he has been alerted to his special guest by Kento Suga, who heads marketing for Niantic in Japan. The steak flows endlessly because the owner is so excited to meet, take a picture with, and get the autograph of the man behind . . . Pokémon GO.

He knows nothing about John's former work on Keyhole and Google Maps and Google Earth. To him—and the rest of the world, it seems—John Hanke is now known simply as the creator of Pokémon GO.

We are in Tokyo because of Ingress, Niantic's first game. The largest event Niantic has created to date took place the day before, and over ten thousand Ingress "agents," as they are called, walked, ran, and cycled through the streets of Tokyo, battling over virtual ownership of the city. These types of events are happening all over the world now. We will produce twenty-six in 2016 and attract hundreds of thousands of Ingress players out into the real world to play a video game.

As big as the Ingress event was, however, it is not what is on everyone's mind, and when I say everyone, by that I mean everyone on Planet Earth.

It has been twelve days since the launch of Pokémon GO in the United States and a handful of other countries, and the world has gone crazy for catching Pokémon. It was the breakout hit of the summer of 2016. The craze has been the subject of skits by Jimmy Fallon, Stephen Colbert, and Jimmy Kimmel. Hillary Clinton attempts a Pokémon GO joke at a campaign stop. There are thousands of press articles and hundreds of millions of social posts; there's even an inset photo of John included on the cover of *People* magazine.

Thousands of people gathered in public parks and enjoyed the outdoors in new ways, and there are player-organized walks and social events. The app was being used more than Twitter or Tinder, and had already broken all-time download records that summer. Upward of one hundred million people were playing, daily. In Dubai, a man stops his car in the middle of the highway and jumps out in pursuit of a rare Pokémon. Two teens climb over a fence in San Diego and fall off of a cliff (they were okay after the accident). Saudi clerics issue a fatwa against the game, declaring it against Islamic religious principles. A reporter disrupts a U.S. State Department briefing to catch a Pokémon. "You're playing the Pokémon thing right there, aren't you?" State Department spokesman John Kirby asked a reporter in the middle of discussing ongoing anti-ISIS war efforts.

“Did you get one?” he asked.

Niantic has not yet released Pokémon GO in Japan, however, and that’s fine with me. I’d prefer to be elsewhere when Pokémon GO launches in the country where the Pokémon phenomenon was born twenty years earlier. There are legitimate fears among Japanese government officials of widespread civil unrest.

“Let me ask you this,” I say to John. “Do you think that the launch of Pokémon GO was possibly even bigger than the launch of Google Maps and Google Earth?” Though he’s been working nonstop for the last two weeks, he is running on adrenaline, just like the rest of the company.

“It’s funny you should mention that,” John says. “Brian sent me a Google Trends graph plotting the popularity of Pokémon GO at launch against the popularity of Google Maps and Google Earth at their launch. GO was three times Google Maps and Google Earth.”

I shake my head in disbelief as I pull a perfect piece of Kobe beef off the hot stones. We are in Roppongi Hills, just down the street from the Mori Tower, where John had only hours earlier met with the CEO of gaming giant Nintendo. When Niantic spun out as a separate company from Google, in October of 2015, Nintendo took the lead investor role in John’s new start-up (the investment talks started because the wife of the Nintendo CEO was a hard-core Ingress player). Over the last twelve days since the launch, Nintendo’s market capitalization has soared 120 percent, from $19 billion to $42 billion. I guessed the meeting had gone well.

“Shit, John, did he try to just buy us outright?” I asked. Evan laughed.

"No, he did not," he quietly said, adding, "but I *was* wondering if he might try to. It wouldn't have been out of the question."

The launch of Pokémon GO had certainly been an unexpected turn of events for John and the Niantic team. The spin-out of Niantic as a separate company from Google eight months earlier had been difficult. It was so sad for me to turn in that Google badge. I loved my eleven years at Google so much. But by 2014, with a new CFO in place at Google, projects that were not deemed core to the Google search mission were being spun out as separate companies. A location-based augmented reality game was not on the list of core projects.

Based on the solid success of Ingress, however, and with Pokémon GO in the pipeline, John was able to raise investment money to create a new Niantic, a company separate from Google (though Google is a follow-on investor in the Niantic 2.0).

After officially spinning out from Google in October of 2015, the company did well, based on our first game, Ingress. I knew what the Niantic investor pitch deck projected, with our typically hopeful revenue and user charts going up and to the right through years one, three, and five. The year-five revenue projections were particularly optimistic. But in the case of Niantic, our revenue estimates would end up being way too *low*. Suffice to say that if we had projected out a fifteen-year chart, we still wouldn't have accurately predicted the revenue Pokémon GO generated that summer of 2016. It broke all download and revenue records and became the fastest-growing application of all time.

With Ingress, John and Niantic had turned the real world into a game board. There are two teams in Ingress, moving in the real world to capture territory. It's a community of gamers drawn back out into their neighborhoods and cities, meeting people and making non-digital friendships. And it gamified the creation of a database of the globe's most interesting locations (portals in Ingress). Those twelve million points collected became the Pokéstops in Pokémon GO.

The idea for Pokémon GO started in 2014, really as a joke. A Google engineer named Tatsuo Nomura created an April Fools' Day joke for Google Maps, overlaying his childhood favorite Pokémon characters onto Google Maps. It was a huge hit globally, and his friend on the Niantic team, Masashi Kawashima, showed it to John, pitching the idea of making it a real-world GPS-based game, like Ingress. It was one of the many next-game ideas John was considering, but he knew what a great fit Pokémon could be. Masashi and John went to Tatsuo and asked him if he might want to make his April Fools' Day joke something more.

Pokémon GO—built on the backs of Ingress learning, data, and technology—has proven to be a true north for an entire new industry. Prior to Pokémon GO, there was no proven use case for augmented reality, or AR as it is called. Now, thanks to Niantic, AR is one of the hottest tech trends around: There are dozens of AR companies in Silicon Valley and elsewhere whose investment pitch decks have a slide or two about Pokémon GO. There will be many more augmented reality games, including (brace yourself) Harry Potter: Wizards Unite, slated for release some time in 2018.

This new, augmented world begins with good maps, with the ability to precisely geo-locate everything around us. For that reason, it's no coincidence that the very first massively successful augmented reality app was built by the same core mapping team that created Keyhole and Google Maps and Google Earth, including Phil Keslin as Niantic's chief technology officer.

Thanks to Google Street View, our friends and former coworkers on the Google Geo team are also well positioned to play a leading role in augmented reality. CEO Sundar Pichai recently announced, at Google I/O 2017, that Google Street View imagery is being used to extract much more than speed limits, street names, and school zones. Now all sorts of objects are being recognized by using computer vision, and are being accurately positioned to become the foundation of new augmented reality services and games.

What does this augmented future look like? Imagine for a moment standing on the University of Texas campus on the corner of 24th Street and Guadalupe. Hold up your phone and point it at that statue. Based on the Google Street View data, and computer-vision-based mapping, your phone recognizes that to be a statue of Congresswoman Barbara Jordan, and a subtle, elegant information bubble immediately floats over her head, showing her name, dates of her life, and key legislative accomplishments. She's outlined in blue, and the statue appears to come alive, and Congresswoman Jordan begins speaking to you, delivering highlights from her keynote speech at the 1976 Democratic National Convention in Madison Square Garden.

Now imagine holding your phone north, along Guadalupe Street or the Drag. You can see a transparent overlay floating above the Hole in the Wall live music bar, a digital marquee showing who is playing tonight with a video clip of each band. As you pan your phone back down Guadalupe, the bench across the street has been previously identified as a bus stop by Google Street View. The bus schedule floats above it, with a countdown clock showing you how many minutes until the next bus. Yelp reviews hover above every restaurant; rates and room availability are suspended above every hotel. This experience may or may not involve a smartphone; it could be spoken into your ear through a barely visible earpiece or viewed on a special pair of sunglasses.

Does this sound like a scene from a sci-fi movie, like *Minority Report* or *Her*? It may sound like the distant future, but it's a world that is coming fast. We are all untethered now: Games are moving off of couches in basements and searches off of monitors in homes—they are moving out into the real world.

Is this a world where people's attentions will be further sucked into their phones? Or will we look up from our phones and appreciate the world around us with new eyes and a greater understanding of the history, architecture, and cultural significance of a place? Will we be more present, with this richer knowledge? Or even more distracted?

And who wins in this new, augmented world, one in which a walk down a city street or a grocery store aisle becomes a hyper-informed, self-guided tour with murals dancing to life and product pitches emanating from wine bottles? My hunch is that it will be those with the best maps, those that are accurately and methodically indexing and positioning every place on the planet. The content will come from the usual sources, but it will need to be perfectly positioned, aligned so that it is inserted into the real world at precisely the right latitude and longitude.

In many ways, this future simply falls on the continuum of the Street View project, the same one Larry started by pointing a camcorder out of his car window in 2002. Maybe this was the plan all along.

AUTHOR'S NOTE

One of my favorite books is Roald Dahl's *Going Solo,* the autobiographical story of the author's time as a Royal Air Force fighter pilot during World War II. Dahl's harrowing, high-wire adventures over the Greek Isles and Northern Africa, including being shot down in Libya, provide the reader a window into the valiant plight of the often-outnumbered British pilots. Dahl's "Shot Down Over Libya" recounting was his first published work—in *The Saturday Evening Post*—and it served to sway the American public's opinion about getting involved in the war (and got Dahl interested in writing).

But these adventures, compelling and entertaining as they are, do not pretend to offer a full compendium on World War II. They are simply Dahl's perspective on the war, his viewpoint out the window of his Hawker Hurricane fighter plane on the violence below. In no way is *Going Solo* a complete accounting of all the battles or all the lives lost, nor is it a full exposition of the technical capabilities of his Hurricane or the German Messerschmitts gunning for him.

In much the same way, I do not purport to say that this book is a full technical compendium of digital mapping innovations at Keyhole, Where2Tech, Google, and beyond. There are important people, transformative technologies, and entire projects whose role and impact are underrepresented and, in some cases, not reported on at all.

These omissions or misattributions are not intentional: This is simply one person's perspective. It is my (mostly nontechnical) window into the mapping revolution that happened all around me, starting with a visit from a friend in 1999. A revolution that continues today.

ACKNOWLEDGMENTS

Late in October of 2014, I returned to the Google Austin office from the ten-year Keyhole acquisition reunion in Mountain View. By this time, I was the old man on the block; I had been at Google longer than any of the five hundred Googlers working in Austin at the time. Inside the walls of Google, the Keyhole acquisition is largely regarded as one of the most—if not the most—successful acquisitions in Google's history. So my Keyholer status gave me more than a bit of street cred inside the company, especially among the young Googlers in Austin. Two of them stopped by my office as I ate my lunch and asked me about the reunion. Feeling nostalgic, I invited the two Googlers in. They sat down on my small black leather couch in my gleaming office with an unobstructed view of the University of Texas tower only six blocks away. I ended up regaling them with more than a few of the Keyhole and early Google Maps stories that had been told at the reunion. After about thirty minutes of this, one of the young Googlers—one whom I had come to know as an entitled, negative, though entertaining, software engineering malcontent—interrupted excitedly: "Dude! You should write a book!" He was wide-eyed and excited in a way that I had never seen him. And so it began.

I should note that I've *begun* many a creative endeavor: I'm great at starting on a project. Finishing a project—not so much.

For that reason, I must start by thanking the one person who would not let me *not* finish this book: my mother-in-law, Robin

Wallace. As an author, playwright, folk musician, seamstress, and painter (and nurse, avid tennis player, international traveler, gourmet cook, political activist), Robin knows well the steps and discipline needed to actually *produce* something creative—not just start it. Over Christmas of 2014, I read her the first few stories that I had written up. She was beyond positive about this book, borderline pestering me with her infectious enthusiasm and encouragement. Without Robin, I can honestly say that this book would be nothing more than a half-written collection of printed-out stories taking up space in my filing cabinet. Thank you, Robin, for being the inspiration to keep going, to keep working at it, and for not letting it die on the shelf.

I knew so little about this whole process, but thankfully Duvall Osteen at Aragi was there to guide me through it all. She was an early advocate for the story and managed to get it into the hands of the right people, at the right time.

I must thank my editor Stephanie Hitchcock at HarperCollins for having the vision of what this book could be and an appreciation of how this technology changed the way we find our way in the world. She was the perfect combination of positive and encouraging when needed, as well as frank and honest when the book was veering off course. It was Stephanie who eventually intervened and told me straight out that I needed help—professional help—to complete the book. Thank you, Sarah Reid at HarperCollins, for project managing this book. I was a rookie in this endeavor, and you helped keep me on task and on schedule. And thank you to the HarperCollins marketing, publicity, and legal teams for getting it out into the world: Beth Silfin, Brian Perrin, Heather Drucker, and Hollis Heimbouch.

I was truly lucky in finding a writing coach, editor, and partner to fundamentally help rewrite the book, S. Kirk Walsh. To use software parlance, I had hacked together a demo; Kirk helped me to create a shippable product. Thank you, Kirk, for saving me from including too many bad jokes and letting me keep the ones that I had to tell. And thank you to author Stephen Harrigan for making the thoughtful introduction to Kirk. And thanks to Chris Hyams of Indeed for reading and providing feedback on the alpha version of the book. Also thanks to Josh Bauer, Bing Gordon, and David Richter for agreeing to read and review advanced copies of the book.

I must thank all of the Keyhole, Where2Tech, and Google former coworkers and friends who helped me piece together the story and fill in the gaps when I wasn't there. I would especially like to thank Chikai Ohazama, Brandon Badger, Wayne Thai, Jens Rasmussen, Luc Vincent, Noah Doyle, David Lorenzini, Rob Painter, Lenette Posada Howard, Ed Ruben, David Kornmann, Mark Aubin, Phil Keslin, Holly Hanke, and Daniel Lederman. I don't pretend to say that this is the whole story of Google Maps and Earth. It's really just my view from where I sat.

Brian McClendon could have easily written this book. With his thorough and organized recordkeeping, he would certainly have created a more complete and technical compendium of the innovations. I appreciate his willingness to share his perspective and, of course, correct me on many dates and facts.

Michael Jones deserves credit for encouraging me not to take sides in the revisionist history battle to claim credit for the success of Google Maps and Google Earth. "We were all a part of creating something great. Give everyone involved the credit.

Share it with everyone," he said. Like I said, Michael is quite likely the smartest person I have ever met. I reminded myself often of his sage advice, especially when writing up the stories of projects where multiple people were involved.

Thanks to Mara Harris at Google for shepherding the book through the proper Google legal reviews.

Thanks to Larry Page and Sergey Brin. Though I have met both of you only in passing, I am in awe of what you created. When you step onto a Southwest plane, you can feel Herb Kelleher, the zany founder. When you set foot in Disneyland, you feel the whimsical personality of Walt Disney. And when you use a Google product, you feel Larry and Sergey's desire to change the world.

And of course, thank you, John Hanke, for encouraging me to write the book. "It's a story that should be told, and you might just be the perfect person to tell it," he once said. He is a great friend, a loyal boss, and a true visionary about the future of technology. I am forever indebted to you, John, and I look forward to writing a book about your next radical transformation.

Thanks to my wonderful extended family. You are a creative and crazy bunch of artists and writers and sailors and lawyers and teachers. As the youngest of eight, I have been truly blessed in the family category.

Lastly, thanks to my wife, Shelley, who was more than along for the whole, sometimes scary, Keyhole and Google ride. You didn't sweat the missed paychecks or expense reports (I did tell you about those, right?). You lived this story with me from those early dark days in Boston to the glorious Google ride and spent many a night in bed listening to me read passages from this book; you gave me those first words of encouragement to get this journey down on paper. Ironically, with you and your uncanny sense of direction, Isabel, Camille, and I never really needed Google Maps on our journey. We really only need you, Shelley. The S in GPS.

ABOUT THE AUTHOR

Bill Kilday's twenty-five-year career in technology and game marketing has centered on maps and augmented reality. He served as marketing director for digital mapping start-up Keyhole, and marketing lead for Google's Geo division during the launch of Google Maps and Google Earth. He is currently VP of marketing for Niantic Inc., a spinout company from Google responsible for GPS-based games Ingress, Pokémon GO, and the upcoming Harry Potter: Wizards Unite. The youngest of eight children, Kilday was born in Houston, Texas, and now lives in Austin, Texas, with his family. He has a terrible sense of direction.